ADVANCED APPROACH TO MITIGATE MAGNETIC FIELDS AND YOUR HEALTH

A Comprehensive Mathematical Modeling for Mitigation of Magnetic Field

A.R. MEMARI

CONTENTS

PREFACE

The main purpose of preparing this book is to share with the readers my 15 years of experience which I have accumulated during my post doctoral research work and in the times that followed.

Attempts have been made to scrutinize each case and provide readers with detailed information and also to demonstrate applicability of the developed procedure.

The author truly hopes that in reference with the scientific reports and epidemiological studies establishing a correlation between exposure to magnetic fields and human's health, this book may be considered as a useful tool to protect health of mankind threatened by exposure to magnetic fields.

Electricity has always contributed tremendous effects on the growth of our societies as well as changing the way mankind has lived. Electricity has gone through a number of changes and has always been challenged to meet the demands. As there is nothing to produce hundred percent advantages, electricity also cannot be of exception.

Even though, in fully developed societies with switching off electricity, life comes to a complete halt, but the hazardous effects of electricity on human's health, especially those who are in direct contact with electricity or operating equipments run by electricity must be deeply investigated and steps must be taken to protect mankind against such fatal threat.

During the past decades, generation of electricity has gone through major changes and consequently, there are numerous methods to develop electricity. Irrespective of how electricity is generated, there is only one way to transport this energy, and that is through transmission line. Electrical power transmission lines are installed between the power plant and a substation. Since it is desired to deliver a large amount of power through a very long distance, transmission normally takes place at high voltage. Redundant lines are provided so that power can be routed from any power plant to any load center. The conductors, which are made of aluminum alloy and reinforced with steel strands and are not provided with insulators require minimum clearance.

When electricity entered market, it used to be delivered at the same voltage as to be used by houses and other electrical equipments, which in turn required different circuits and the distance between the power plant and the consumers was kept restricted.

In order to increase the distance between towers, which results in reducing the cost of transmission, clad steel wires and high towers are used. The number of towers per kilometer distance can be reduced to as few as 6 towers. The longest high voltage transmission line, which is installed in Republic of Congo, has a length of 1700 kilometers.

Engineers are always concerned about the power loss in transmission line. This loss, which is dissipated as heat due to the resistance, is proportional to the surface area of the transmission line conductors. Consequently, the smaller the surface area, the lower will be the loss due to heat dissipation. However, at very high voltages, corona discharge losses become very large. With high voltage transmission line, the voltage is stepped up at the generating station and then stepped down to the voltage needed by the distribution network. Such process increases the transmission efficiency.

Voltages lower than 110 KV are considered as sub-transmission voltages, whereas voltages above 230 KV are known as extra high voltage. A network of transmission lines, substations and power plants is known as transmission grid.

Due to high cost of generation of electricity, a strong possibility exists to import the required extra power in places where consumption of electricity is variable (due to hot summer and cold winter).

A.R. Memari, *Ph.D., P. Eng.*

ABOUT THE AUTHOR

The author is a holder of Ph.D. degree in High Voltage Engineering and has been engaged in Post Doctoral Research work at the University of Toronto, Toronto, Canada. Dr. Memari has published numerous scientific papers in the area of magnetic field associated with high voltage transmission line. His research findings have been utilized worldwide. In addition to American inventors who have utilized his research findings for their invention, Japanese Scientists have also proven the practical implementation of his research works. Dr. Memari has also published a book related to Law and Ethics.

Dr. Memari is a full member of Association of Professional Engineers Ontario (Canada) and a Professional Member of Ontario Society of Professional Engineers. He is also in possession of a patent at Canadian Intellectual Property Office.

Title of Invention: "Implementation of Capillary in Generation of Electricity".

Dr. Memari's second patent, which deals with mitigation of magnetic field contributed by household appliances, will be filed in the near future.

CHAPTER 1

Exposure to Electromagnetic Fields

Abstract: This chapter deals with the epidemiological reports, which associates exposure to magnetic field and the risk of fatal diseases such as brain cancer, leukemia, breast cancer, depression and miscarriage. In addition to exposure to magnetic field associated with high voltage transmission lines, possible relationship between health hazard and exposure to residential magnetic field such as electric blanket, hair dryer, toaster and many more have also been discussed.

Occupational and non-occupational exposure to magnetic field and how an office worker's health may be in danger by sitting near to some office equipment have been studied.

Finally, a relation between central nervous system cancer and environmental exposure has been established. This study also establishes a possible relation between modern equipments and heart attack.

1.1. Electric and Magnetic Fields

Electric and magnetic fields are products of power lines, industrial machineries and electrical equipments, such as computers, electric blankets, electric clocks, table lamps, hair dryers, televisions, microwaves and many more.

Electromagnetic field comprises two components known as electric field and magnetic field, which are investigated separately in low-frequency rate and since electromagnetic does not have enough energy to rupture molecular bonds, it is known as non- ionizing radiation.

Electric fields are generated by voltage and its strength increases as the voltage is increased. Its unit is volts per meter and could be present even when the electrical equipment is turned off. Walls and trees can easily shield electric fields. Electric field can easily cause electrically charged aerosols to oscillate and since our body is a good conductor, electric field around our head is increased by a factor of 18.

Magnetic fields are products of flow of current in any conductor. It is directly proportional to the current flowing through the conductor, but it is inversely proportional to the distance from the conductor carrying the current. Magnetic field has a unit of gauss or tesla. Since magnetic field is a direct product of current, it can be present as long as current is flowing. Consequently, it vanishes the moment the electrical equipment is turned off.

It is only magnetic field that has been found responsible to cause cancer, leukemia, and other health hazard in children and adults.

Health problems related to exposure to magnetic field began in USSR for the first time in 1960. In the early stage, the researchers concentrated on electric field because high voltage transmission lines are well capable to produce more current in our body rather than magnetic field. But the research could not find any evidence relating human's health problem to electric field, they therefore focused on magnetic fields.

It was in 1979 when for the first time researchers established a relationship between leukemia, tumors of the nervous system in the children living near high voltage transmission lines. These researchers noticed that these children had doubled or tripled their risk of developing these diseases.

Hazardous effects of magnetic fields have become more visible in recent years and since more data and evidence are available, public and politicians are more concerned about this invisible radiation generated by transmission lines, electric wiring, household appliances and substations.

ELF magnetic field can penetrate human body, resulting in malfunctioning of nerve and muscle cells and therefore, in order to protect health of general public against this invisible radiation, an international limit must be established.

A report by National Institute of Environmental Health Science suggests that due to insufficient evidence associating exposure to magnetic field with health hazard, an aggressive regulatory action cannot be taken.

It is a misleading to use job title as a surrogate for exposure in occupational exposure to magnetic field. There exists a remarkable difference between the field survey exposure and the actual level of magnetic field to which a person may actually be exposed. Even though hospital workers are non-electrical workers, but they are in contact with equipments such as diagnostic, monitoring equipment, visual display terminal, photocopier and many more, which are well capable of producing high level of magnetic field.

The instruments used by non- electrical workers can expose them to a high level of magnetic field. For example, an electric drill is well capable of producing magnetic field as high as 5000 mG close to its handle, while many electric utility workers could be exposed to magnetic field of 2 mG. Even though, office workers are generally exposed to low level of magnetic field, but the overall exposure to magnetic fields should not be ignored. It is estimated that an office worker may be exposed to a magnetic fields as high as 100 mG as the officer might be sitting near to some office equipment, such as photocopier, which may last for a long period of time everyday.

The Institute of Electrical and Electronics Engineers also reports of association of exposure to magnetic field generated by high voltage transmission lines and household appliances and human's health.

Threat relating magnetic field to health problem in the general public has encouraged researchers and scientists to conduct further investigations.

Even though the available data is not sufficient to truly believe that magnetic field stands as a major threat to human health, there is also no enough evidence to believe that magnetic field does not stand as a major threat. In the past decades scientists and researchers of numerous countries have been deeply involved in collecting more data and evidence regarding association of several types of cancer with exposure to magnetic field.

There are still disagreements among researchers about level of safety of magnetic field and whether magnetic field causes diseases or promotes them.

Effect of magnetic field can be detected at low level, but it may lose its effect as the level is raised, but this effect reappears, as the level is moved to a higher level. Consequently, level of safety of magnetic field has become a ground of further investigation and research.

For the past decades, scientists have become more concerned about possible relationship between health hazards in particular, cancer, leukemia, brain tumor, depression and suicide, and exposure to residential magnetic field as well as that of the high voltage transmission lines.

Even though, numerous studies in the past established a relationship between magnetic field exposure and childhood leukemia, but collaboration between National Cancer Institute and Children Cancer Group, resulted in finding little evidence linking exposure to magnetic field to development of acute lymphoblastic leukemia in children under age of 15. Also, these studies have not been able to establish a relationship between exposure to magnetic field and children brain tumor.

National Cancer Institute reports that children living in houses with high level of magnetic field did not have an increased risk of childhood acute lymphoblastic leukemia. National Cancer Institute also indicates that children living near high voltage transmission lines were not at greater risk of leukemia. However some studies indicate that magnetic field above 0.4 μT increases risk of childhood leukemia, though some researchers put this value at 0.3 μT. Surprisingly, some other research institutes put the maximum allowable level of exposure at 100 μT and even at 1.6 mT, which are 250 and 4000 times higher than the level above which the risk of childhood leukemia has been observed to be doubled.

Some evidence shows that exposure to magnetic field of as low as 0.1 μT could stand harmful to our body. This level of magnetic field can easily be found in vicinity of high voltage and low voltage transmission lines. Perry *et al.* suggests a 40% increase in suicide risk above 0.1 μT.

In UK an advisory group reports that between 2 and 4 out of 500 cases of childhood leukemia per year could be due to elevated magnetic fields.

In a research conducted about effect of magnetic field of distribution systems, showed that children living near such systems have doubled their risk of cancer death, and some other researchers have found association between childhood leukemia and household exposure to magnetic field.

Contrary to the above findings, other scientists conclude that the danger may have been exaggerated.

UK government has recently issued a moratorium on constructing houses and schools within 60 meters of the existing high voltage transmission lines. It may soon (in some countries) be required for public to install special devices to check on their household magnetic field exposure.

Some European countries, such as Switzerland, are planning to rewire the overheads to reduce the magnetic fields, and Australia have established a precautionary limit for magnetic field, with which the World Health Organization has also agreed and has recommended further research on sources of exposure to low-level magnetic field and their link to health hazard such as cancer, depression, suicide, reproductive problems, developmental disorders, immunological changes and neurological diseases.

Researchers at University of Toronto and the Hospital for Sick Children have found an association between magnetic field exposures in residences with the risk of developing childhood leukemia. According to theses findings, children who were exposed to high level of magnetic fields in their houses were more likely to develop leukemia. This number jumped to more than four times when other factors such as power consumption, child's medical history and environmental exposures were taken into consideration. This risk was higher for children less than six years of age.

However, some researchers at the University of Toronto still are not convinced that magnetic field causes cancer nor laboratory research support such claim. Some believe that there is no enough biological explanation for how such exposure may work.

Study conducted by Children Cancer Research Group and Transco, report that they could find 70 percent increased risk of leukemia in children living in the radius of 200 meters from high voltage transmission lines. This percentage dropped to 20 percent for those who were living within distance of 200 meters to 600 meters from the transmission lines. Association of miscarriage, suicidal tendencies and adult leukemia with exposure to magnetic field has also been reported.

A United Kingdom newspaper has reported that in research work conducted on pregnant women, 14 percent of pregnant women who had lived in the boundary of 25 meters from high voltage transmission lines had miscarried. Whereas this percentage dropped to only 3.5 percent for the pregnant women who were living further away from the power lines. 27 percent of the women living in the vicinity of the power lines reported of depression, compared with 13 percent living further away.

Among the women participated in the research work, 63 percent of those who were living in a boundary of 25 meters from the power lines, complained of regular headache, but only 39 percent of these women who were living outside this boundary had suffered from headache.

The research work conducted in Canada reports that pregnant ladies exposed to occupational high level of magnetic field has caused childhood leukemia.

It has also been reported that dressmakers are at a very high risk of developing Alzheimer's disease due to the exposure to high level of magnetic fields produced by their sewing machines, which is estimated to be three times higher than that of power lines.

Amyotrophic lateral sclerosis is a fatal neurodegenerative disease. Recent studies establish a correlation between occupational exposure to ELF electromagnetic fields and their effects on central nervous, immune and musculoskeletal systems.

Effect of ELF magnetic field on brain and nervous system of mammalian has been observed, which includes alteration in the EEG pattern of several animals such as rats, pigs and monkeys.

Mice entered a state of physically and mentally inactive, when it was exposed to 60 Hz electrical fields above 100 kv / m.

The research works conducted by USSR researchers in 1970's established a relationship between 50-Hz electric field and symptoms such as headache, fatigue, increased irritability and reduced performance and activity levels as well as sexual weakness, but some American scientists have not been able to show a significant increase in the above mentioned symptoms.

Epidemiological findings report of association of exposure to occupational magnetic field and the risk of amyotrophic lateral sclerosis, which is characterized by a progressive degeneration of the cortical and spinal motor nerve cells. Even though more investigations have been recommended, but a correlation between occupational exposure to magnetic field and dementia has also been reported.

Exposure to occupational magnetic field and neurodegenerative diseases such as Alzheimer, Parkinson indicates an increase risk of dying from Alzheimer due to exposure to magnetic field exceeding 0.5 µT. There are controversial reports linking risk of dying from Amyotrophic lateral sclerosis to exposure to magnetic fields.

Even though, the past studies could not establish a relationship between Parkinson disease and exposure to magnetic fields, but recent investigations report of small possibility between Parkinson disease and exposure to ELF magnetic field. Since, welders are not only exposed to magnetic field, but also exposed to metallic fumes and other chemicals, are at higher risk of developing this diseases.

Studies conducted on nuerodegenerative and exposure to magnetic field during period of five years (2000 – 2005) by different scientists were heterogeneous. In these types of studies, not only data sources and analytical methods varied, but also different methods to evaluate exposure to magnetic field were adopted. These studies were mainly based on the occupational exposure, but exposure to non- occupational magnetic fields were hardly investigated. Elements such as aluminum, lead and manganese are also capable of generating neurodegenrative diseases.

1.2. Brain Cancer

Even though the true cause of brain tumor is not well known, but an association of genetic factor and environmental exposures with brain tumor has been established. Certain syndromes are found to be responsible to develop brain tumor, however these syndromes are rare. Children whose parents are suffering from brain cancer are likely to have a slight increased risk of developing brain cancer. Effect of magnetic field on central nervous system cancer has established a possibility that cellular phone as well as other sources of magnetic field may cause cancer.

Even though high level of radiation from radiotherapy can surely cause brain tumor, but there is no strong evidence to associate magnetic field with brain tumor.
Insufficient information available hinders scientists to believe that magnetic field causes cancer. Subsequently, further studies are being conducted to learn more about effects of magnetic field exposure on human's health.

Some researchers report that exposure to magnetic field constitutes effect on short-term memory.

Occupational exposure to magnetic field has been a ground of interest to the scientists in Canada and the research work conducted on utility workers, power station operators, phone line workers showed an increased risk of developing brain tumor, leukemia and male breast cancer.

Researchers at the University of Pittsburgh have found that exposure to high level of current at work place in an aluminum plant has caused death, resulting from leukemia and lymphoma.

1.3. Breast Cancer

The environmental magnetic fields effects on breast cancer have also been of immense interest to the researchers and scientists of this field. Such an interest resulted in finding a relationship between 60 Hz magnetic fields and an increased risk of breast cancer. As our societies are becoming more industrialized, risk of developing breast cancer among women is also increasing. Modernization of our society requires more electricity and other devices such as phones, computers, hair dryers etc., resulting in generating more magnetic fields.

Higher rate of breast cancer among women in industrialized countries in comparison with less industrialized countries has established a ground for the researchers and scientists of this field to associate magnetic field with increased risk of developing this diseases.

Experiments conducted on animals indicated that melatonin has a tremendous effect on breast cancer, but it is difficult to gather evidence on human being. Melatonin is very effective to develop several types of cancers in animals, including breast cancer.

Some scientists, in early 80s established an association between melatonin suppression and magnetic field. These scientists reported that as level of melatonin is reduced, risk of developing breast cancer is increased. Melatonin level is increased during night but decreases during daytime.

Melatonin is a hormone produced in response to a lack of light by pineal gland, which is located at the base of the brain. Consequently, growth of melatonin can be affected by lights during the nighttime, which in turn may influence behaviour, mood, or even immune systems.

It is believed that melatonin levels are inversely proportion to estrogen levels. Subsequently, when melatonin levels are low, estrogen levels are high or vice versa. Therefore, if nighttime light or magnetic fields suppress the normal nocturnal rise in melatonin, estrogen levels would subsequently be increased. Since increased levels of estrogen are hypothesized to increase the risk of breast cancer, this suppression of melatonin by either light or magnetic field could possibly increase the risk of breast cancer.

It has also been reported that 50 Hz rotating polarized magnetic field above 1.4 µT suppressed both serum and pineal gland melatonin. Such fields are commonly found in the vicinity of power lines. The type of magnetic fields used in some laboratories animal experiments are of linear type, which go through zero twice every cycle. Three - phase power line produces magnetic field vector, which rotates in space, having a minor and major axes known as elliptically polarized. Consequently, this type of field contributes different effect on polar molecules than linear fields.

60 Hz magnetic field can influence melatonin's natural oncostatic action at cellular level. Therefore a 12 mG 60 Hz magnetic field that can very often be found at home and work place can block melatonin's natural growth inhibition of estrogen-positive MCF-7 human breast cancer cell growth in cell culture. This investigation concludes that a 12 mG magnetic field prevents the oncostatic action of both melatonin and tamoxifen.

Breast cancer stands second to lung cancer for mortality among women. The rate of developing breast cancer is almost 12.5 percent and this rate is highest in North America and Northern Europe and lowest in Asia and Africa.

A comparison performed by researchers at the University of North Carolina between female electrical workers and non-electrical workers indicated that the first group was 40 percent more at risk of dying from breast cancer.

Large number of women in Marine are diagnosed with breast cancer for which high level of magnetic field exposure is considered to be the contributing factor.

Some researchers strongly believe that high level of magnetic field exposure in Marin is causing and promoting breast cancer and even though, some cannot find a plausible link, but a strong belief exists that, if magnetic field does not directly cause cancer, but it accelerates the disease that was initially caused by other sources.

Some investigations show that there was no increased risk of cancer among people living for a considerable period of time at a distance of 100 meters from high voltage transmission lines. In some other reports this distance is reduced to 50 meters.

It has been reported that women with blindness are less at risk of developing breast cancer than those who are not.

A laboratory work conducted on chicken brain cell showed that exposure to magnetic field can slow the outflow of calcium in it. Magnetic field exposure can also cause speeding up the copying of DNA strands. Effect of magnetic field exposure has also been observed on disrupting lipid membranes of the cell.

1.4. Appliances

In addition to exposure to magnetic field of high voltage transmission lines, researchers have also been conducting studies mostly in the form of epidemiological to verify the possible relationship between residential magnetic fields exposures and diseases such as leukemia, breast cancer, depression and many more in children and adults.

Household appliances such as videos, electric blankets, refrigerators, computers and many more are well applicable of producing magnetic field.

Appliances such as refrigerators that can produce high level of magnetic fields should be kept in a place where they are not in frequent use by the residents.

Researchers believe that magnetic field produced by a toaster could be more harmful to the public health than that of a transmission line.

Recently, some governments have advised their people to mitigate their exposure to magnetic field from common household appliances.

In the recent studies, use of household appliances and childhood acute lymphoblastic leukemia has been scrutinized. In this type of studies, use of electrical equipments by mother during her pregnancy and then by the same child (after birth) has been thoroughly investigated.

These studies found that since each appliance is in use for a short time and the level of magnetic field generated by this appliance mitigates as the person is standing away from it, therefore the overall exposure of the person to household appliances is less than that of power lines.

Epidemiological studies have established a relationship between central nervous system cancer and environmental exposures. Some scientists establish a possibility that cellular phone may cause cancer.

Researchers at the University of North Carolina have found that fatal heart attack is common among utility workers exposed to high level of magnetic field. Researches on this type of diseases showed that more than four hundred utility workers who have been working for more than thirty years died from heart attack.

It is believed that 60 Hz current reduces heart rate and the report indicates that exposure to 60 Hz magnetic field among utility workers has even doubled (in some cases) their risk of suffering from heart attack.

CHAPTER 2

Fundamental of Magnetic Field

Abstract: Electricity has gone through lots of changes since it has been in use. But, the hazardous effects of high voltage transmission line magnetic field have always been of deep concern to scientists. In this chapter, a mathematical modeling has been established, by which a hundred percent mitigation at any point of consideration would be achievable. The developed equations are also capable of producing simultaneous reduction of magnetic field at other points within the center of the right-of-way.

First, a procedure to obtain the total magnetic field contributed by the three- phase high voltage transmission line has been established and then the angular frequency at which maximum magnetic field occurs is developed.

In order to achieve the mitigation, an auxiliary mitigating loop has been implemented and the required equation to calculate the mitigating magnetic field is developed. The optimum value of this loop impedance is set. Since the designed auxiliary mitigating loop is of passive type, no especial feedback equipments are needed to compensate for the changes of the load currents.

In order to illuminate the capability of the developed equations a flat configuration of 230 KV transmission line has been utilized.

2.1. Magnetic Field

When a current carrying conductor is placed at point (x_i, y_i), it produces magnetic field at point (x_j, y_j), which is directly proportional to current $\bar{I}$, but it is inversely proportional to the distance r_{ij} as shown in Equation (1).

$$\vec{H}_{ij} = \Sigma \left(\frac{\bar{I}}{2 * \pi * r_{ij}} \right) \Phi_{ij} \quad (1)$$

Where:

$\Phi_{ij} = \left(\cos(\alpha) + \sin(\alpha) \right) (2)$ and r_{ij} is the distance from the conductor to the point of consideration. α is the angle that r_{ij} makes with the vertical line h, as shown in Figure 1

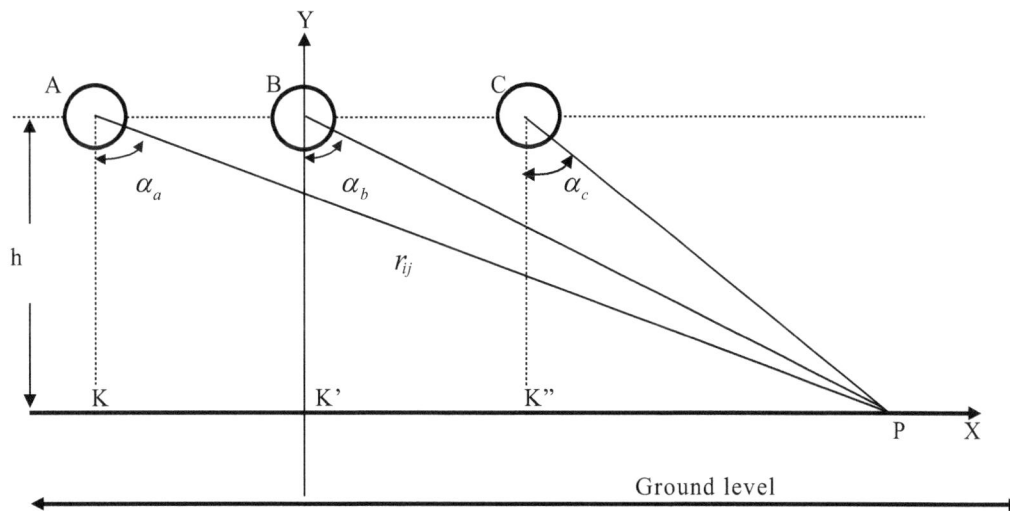

Figure 1. Position of the three phases with respect to an arbitrary point.

Since current $\vec{I}$ is sinusoidal, the generated magnetic field also varies sinusoidally in such manner that during positive half cycle its direction coincides with that of the directional vector. It obtains an opposite direction during the negative half a cycle.

Equation (1) would have a trajectory of a circle if the directional vector was not included. At distances beyond hundred meters from the transmission line, the effect of earth return current must be considered. Such consideration, consequently requires a correction factor to be added to Equation (1)

Since the currents in the transmission lines vary sinusoidally, the magnetic fields produced by these currents, obviously, vary sinusoidally. The resultant magnetic field vector is the vector sum of the magnetic field generated by each line, as expressed by Equation (3). Sinusoidal variation of currents with respect to time and having a relative phase displacement with respect to each other, the corresponding magnetic fields at the point of consideration add up vectorially to produce the resultant magnetic field, $\vec{H}_T$.

$$\vec{H}_T = \vec{H}_A + \vec{H}_B + \vec{H}_C \quad (3)$$

Where $\vec{H}_A, \vec{H}_B$ and $\vec{H}_C$ are the magnetic fields produced by phase A, phase B and phase C respectively. $\vec{H}_A$, $\vec{H}_B$ and $\vec{H}_C$ variations are sinusoidal along their own directions, but due to phase displacement, there is a change with time not only in the magnitude of the resultant vector $\vec{H}_T$ but also in its orientation. These changes constitute the tip of magnetic field vector $\vec{H}_T$ to establish an ellipse, when angular frequency is allowed to vary over 360°. Magnitudes and orientations of the major and minor axes are determined by finding maximum and minimum points. The equation of this ellipse can be expressed in terms of two orthogonal vectors.

The three sinusoidally varying currents are expressed as $\vec{I}_A$, $\vec{I}_B$ and $\vec{I}_C$ having a phase difference of 120° and are depicted below.

$$\vec{I}_A = I\angle\theta = I[\cos(\omega t + \theta) + j\sin(\omega t + \theta)]$$
$$\vec{I}_B = I\angle(\theta - 120°) = I[\cos(\omega t + \theta - 120°) + j\sin(\omega t + \theta - 120°)]$$
$$\vec{I}_C = I\angle(\theta + 120°) = I[\cos(\omega t + \theta + 120°) + j\sin(\omega t + \theta + 120°)] \quad (4)$$

Where ωt is angular frequency. The resultant magnetic field produced by the high voltage transmission line at the point of consideration has four components, two real components of X and Y and two imaginary components of X and Y as shown in Equation (5)

$$\vec{H}_T = (H_{rx} + jH_{ix})\vec{i}_x + (H_{ry} + jH_{iy})\vec{i}_y \quad (5)$$

The two real components of X and Y, H_{rx} and H_{ry}, are responsible to establish the orientation in space. Variation of angular frequency over one complete cycle establishes the locus of the magnetic field vector.

The three currents $\bar{I}_A, \bar{I}_B, \bar{I}_C$ are responsible to generate the induced voltage in the mitigating loop. Consequently, the mitigating magnetic field is always proportional to the three-phase current. Therefore, any changes in the value of these currents not only affect the unmitigated magnetic field, but also the mitigating field. This effect occurs in such proportionality that the resultant mitigated magnetic field remains unchanged.

Let us investigate the magnetic field generated by phase A. Since the real component of the current is responsible to establish the orientation in space;

$$\bar{H}_A = \left[\frac{I \cos(\omega t + \theta)}{2\pi R_A} \right] [\cos(\alpha_a) + i \sin(\alpha_a)] (6)$$

where; I is magnitude of the current in phase A. R_A is the orthogonal distance from center of phase A to the point of consideration. From Figure 1;

$$\cos(\alpha_a) = \frac{-h}{R_A}$$

$$\sin(\alpha_a) = \frac{PK}{R_A}$$

$$R_A = sqrt\left((PK)^2 + (h)^2 \right) (7)$$

The magnetic field produced by phase B is given by (8).

$$\bar{H}_B = \left[\frac{I \cos(\omega t + \theta - 120°)}{2\pi R_B} \right] [\cos(\alpha_b) + i \sin(\alpha_b)] (8)$$

From Figure 1;

$$\cos(\alpha_b) = \frac{-h}{R_B}$$

$$\sin(\alpha_b) = \frac{PK'}{R_B}$$

$$R_B = sqrt\left((PK')^2 + (h)^2 \right)$$

Similarly

$$\bar{H}_C = \left[\frac{I\cos(\omega t + \theta + 120°)}{2\pi R_C}\right][\cos(\alpha_c) + i\sin(\alpha_c)]\,(9)$$

$$R_C = sqrt\left((PK'')^2 + (h)^2\right)$$

$$\cos(\alpha_c) = \frac{-h}{R_C}$$

$$\sin(\alpha_c) = \frac{PK''}{R_C}$$

Implementing Equation (3), the total magnetic field produced by the transmission line at any point is given by;

$$\bar{H}_T = \left[\frac{I\cos(\omega t + \theta)}{2\pi R_A}\right][\cos(\alpha_a) + i\sin(\alpha_a)] + \left[\frac{I\cos(\omega t + \theta - 120°)}{2\pi R_B}\right][\cos(\alpha_b) + i\sin(\alpha_b)]$$
$$+ \left[\frac{I\cos(\omega t + \theta + 120°)}{2\pi R_C}\right][\cos(\alpha_c) + i\sin(\alpha_c)]\,(10)$$

In order to establish the locus of the total magnetic field vector $\bar{H}_T$, angular frequency is allowed to vary over one complete cycle of 360°.

2.2. Calculation of Maximum Field

In reality there are four components forming the magnetic field, real and imaginary of X-component and real and imaginary of Y-component, but only the real component of X and real component of Y are responsible in forming the locus of the magnetic field. Therefore, in order to determine the value of angular frequency ωt at which maximum magnetic field occurs, the real value of X-component is vectorially added up to the real value of Y-component.

Derivative of Equation (10) with respect to ωt and equating it to zero results in the value of angular frequency at which maximum magnetic field occurs.

$$\frac{dH_T}{d\omega t} = \frac{I}{2\pi}\left[\begin{array}{l}\left(\dfrac{-\sin(\omega t)}{R_A}\right)(\cos(\alpha_a) + \sin(\alpha_a)) + \left(\dfrac{0.5\sin(\omega t) + 0.866\cos(\omega t)}{R_B}\right)(\cos(\alpha_b) + \sin(\alpha_b)) + \\[2ex] \left(\dfrac{0.5\sin(\omega t) - 0.866\cos(\omega t)}{R_C}\right)(\cos(\alpha_c) + \sin(\alpha_c))\end{array}\right]$$

$$\frac{dH_T}{d\omega t} = 0$$

Therefore;

$$\omega t = a\tan\left[\frac{\dfrac{0.866}{R_C}\left(\cos(\alpha_c)+i\sin(\alpha_c)\right)-\dfrac{0.866}{R_B}\left(\cos(\alpha_b)+i\sin(\alpha_b)\right)}{\dfrac{0.5}{R_C}\left(\cos(\alpha_c)+i\sin(\alpha_c)\right)+\dfrac{0.5}{R_B}\left(\cos(\alpha_b)+i\sin(\alpha_b)\right)-\dfrac{1}{R_A}\left(\cos(\alpha_a)+i\sin(\alpha_a)\right)}\right] (11)$$

Even though a possibility exits to implement the available equations and calculate minor and major axes of an ellipse shaping the trajectory of the magnetic field, but Mat Lab programming which is used to calculate the magnetic field is well applicable to draw the corresponding trajectory and henceforth, no further calculations would be required.

2.3. Mitigating Loop

In order to establish process of reducing the magnetic field produced by the three phases of a transmission line, a loop known as auxiliary mitigating loop is placed either underneath or above the two outer phases of the transmission line. This loop comprises two conductors. In order to modify the impedance of this loop, a series capacitor is included. As already mentioned, the three phase currents induce a voltage in the mitigating loop as shown in Equation (26). Knowing the impedance of this loop, the loop voltage is utilized to calculate the mitigating current in the auxiliary loop. The unique merit of this approach is that the mitigating current is generated by the loop voltage, which is induced by the three-phase current of the transmission line. Consequently, the mitigating current is always proportional to the load current. Therefore, any changes of the load current would have an equal effect on unmitigated magnetic fields. Unlike active circuit that requires a feedback mechanism, no such devices would be needed in case of a passive circuit. As a result the mitigated magnetic field remains unchanged for any value of the load current. Consequently, at all the time mitigating magnetic field is proportional to the unmitigated magnetic field.

Similar to unmitigated magnetic field, as was shown in Equation (5), the mitigating magnetic field also comprises four different components, real and imaginary of X-component and real and imaginary of Y-component as has been expressed by Equation (12).

$$\vec{H}_m = \left(H_{mrx}+iH_{mix}\right)\vec{u}_x + \left(H_{mry}+iH_{miy}\right)\vec{u}_y (12)$$

In Equation (12), $\vec{H}_m$ is the total mitigating magnetic field. The two terms in the first bracket on the right hand side are real and imaginary values of mitigating magnetic field in X-direction. The real and imaginary of Y-component of this magnetic field are shown in the second bracket on the right hand side. $\vec{u}_x$ and $\vec{u}_y$ are the unit vectors in X and Y directions respectively.

As Figure 2 shows, conductor M_1 of the mitigating loop is placed beneath phase A of the power line at a distance of r_1 meters from the point of consideration. As this Figure shows, r_1 makes an angle of β_1 with the vertical line passing through center of phase A. Similarly, conductor M_2 of the auxiliary mitigating loop is installed underneath phase C of the power line. r_2, which makes an angle of β_2 with the vertical line, is the orthogonal distance from the center of conductor M_2 to the point of consideration.

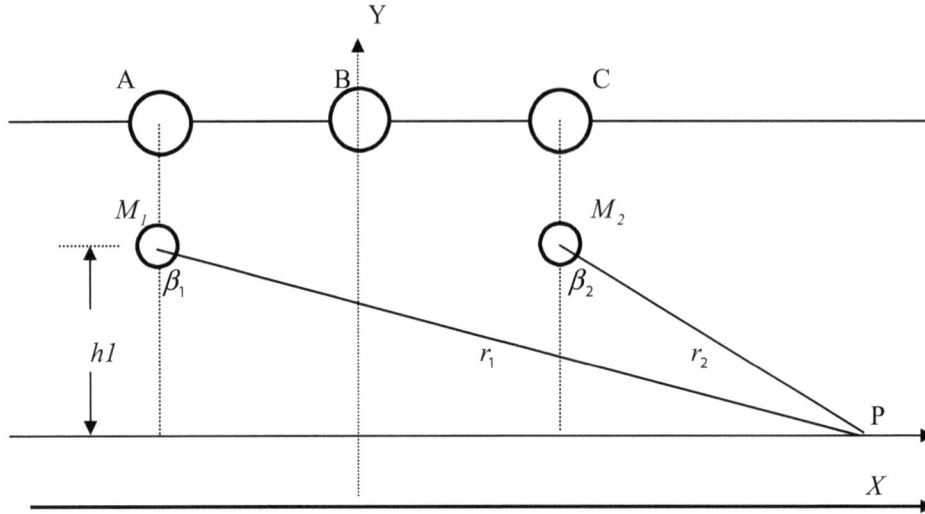

Figure 2. Position of auxiliary mitigating loop with respect to the point of consideration.

From Equation (12), it can be expressed that;

$$\bar{H}_m = \left[\frac{I_m \cos(\omega t + \theta + \gamma) + iI_m \sin(\omega t + \theta + \gamma)}{2\pi r_1}\right](\cos(\beta_1) + i\sin(\beta_1)) -$$

$$\left[\frac{I_m \cos(\omega t + \theta + \gamma) + iI_m \sin(\omega t + \theta + \gamma)}{2\pi r_2}\right](\cos(\beta_2) + i\sin(\beta_2)) (13)$$

Where:

$$\cos(\beta_1) = \frac{-h_1}{r_1}$$

$$\sin(\beta_1) = \frac{-PK}{r_1}$$

$$\cos(\beta_2) = \frac{-h_1}{r_2}$$

$$\sin(\beta_2) = \frac{-PK''}{r_2}$$

With;

$$r_1 = sqrt\left(\left(PK\right)^2 + \left(h_1\right)^2\right)$$

$$r_2 = sqrt\left(\left(PK''\right)^2 + \left(h_1\right)^2\right)$$

After some manipulations, the four components of the mitigating magnetic field could be written as given by Equation (14).

$$H_{mrx} = \frac{I_m}{2\pi}\left[\cos\left(\omega t+\theta+\gamma\right)\left(\frac{\cos\left(\beta_1\right)}{r_1}\right) - \cos\left(\omega t+\theta+\gamma\right)\left(\frac{\cos\left(\beta_2\right)}{r_2}\right)\right]$$

$$H_{mry} = \frac{I_m}{2\pi}\left[\cos\left(\omega t+\theta+\gamma\right)\left(\frac{\sin\left(\beta_1\right)}{r_1}\right) - \cos\left(\omega t+\theta+\gamma\right)\left(\frac{\sin\left(\beta_2\right)}{r_2}\right)\right]\,(14)$$

$$H_{miy} = \frac{I_m}{2\pi}\left[\sin\left(\omega t+\theta+\gamma\right)\left(\frac{\sin\left(\beta_1\right)}{r_1}\right) - \sin\left(\omega t+\theta+\gamma\right)\left(\frac{\sin\left(\beta_2\right)}{r_2}\right)\right]$$

$$H_{mix} = \frac{I_m}{2\pi}\left[\sin\left(\omega t+\theta+\gamma\right)\left(\frac{\cos\left(\beta_1\right)}{r_1}\right) - \sin\left(\omega t+\theta+\gamma\right)\left(\frac{\cos\left(\beta_2\right)}{r_2}\right)\right]$$

The mitigating current I_m Amps creates a magnetic field whose direction is opposite to that of unmitigated magnetic field of the three- phase transmission line. In order to achieve a hundred per cent reduction of magnetic field at any desired location and also to simultaneously obtain a noticeable reduction of magnetic fields at other locations, maximum mitigating magnetic field must be equal in magnitude but opposite in orientation with unmitigated magnetic field.

Therefore, the real components of the mitigating magnetic field, H_{mrx} and H_{mry}, given by Equation (14) must satisfy Equation (15)

$$\left|H_{mrx}\bar{u}_x + H_{mry}\bar{u}_y\right|_{max} = -\left|H_{rx}\bar{u}_x + H_{ry}\bar{u}_y\right|_{max}\,(15)$$

As the first step to fulfill this achievement, from Equation (11) the value of ωt is calculated. Substituting the corresponding value of ωt in Equation (10) maximum value of unmitigated magnetic field is achieved. The next step would be to calculate value of γ angle. Take derivative of Equation (16) with respect to ωt and then equate it to zero.

Equation (13) can be re-written as;

$$\bar{H}_m = \frac{I_m}{2\pi}\left[\cos\left(\omega t+\gamma\right)\left(\frac{\cos\left(\beta_1\right)}{r_1} + \frac{i\sin\left(\beta_1\right)}{r_1} - \frac{\cos\left(\beta_2\right)}{r_2} - \frac{i\sin\left(\beta_2\right)}{r_2}\right)\right]\,(16)$$

$$\frac{dH_m}{d\omega t} = \left[-\sin(\omega t)\cos(\gamma) - \cos(\omega t)\sin(\gamma) \right]$$

Let $\dfrac{dH_m}{d\omega t} = 0$

Therefore;

$$-\sin(\omega t)\cos(\gamma) = \cos(\omega t)\sin(\gamma)$$

Dividing both sides by $\cos(\omega t)$ and then by $\cos(\gamma)$, therefore

$$-\tan(\omega t) = \tan(\gamma)$$

Consequently,

$$\gamma = -\omega t$$

2.4. Calculation of Mitigating Loop Impedance

As has already been mentioned, in order to achieve a hundred per cent mitigation at any desired location, the unmitigated and mitigating magnetic field must be equal in magnitude to each other, but their orientations must be in the opposite directions.

Therefore;

$$\frac{I}{2\pi}\left[\begin{array}{l} \left(\dfrac{\cos(\omega t)}{R_A}\right)\left(\cos(\alpha_a)+i\sin(\alpha_a)\right) + \left(\dfrac{\cos(\omega t - 120°)}{R_B}\right)\left(\cos(\alpha_b)+i\sin(\alpha_b)\right) \\ + \left(\dfrac{\cos(\omega t + 120°)}{R_C}\right)\left(\cos(\alpha_c)+i\sin(\alpha_c)\right) \end{array} \right] = -$$

$$\left[\frac{I_m \cos(\omega t + \theta + \gamma)}{2\pi r_1}\right]\left(\cos(\beta_1)+i\sin(\beta_1)\right) + \left[\frac{I_m \cos(\omega t + \theta + \gamma)}{2\pi r_2}\right]\left(\cos(\beta_2)+i\sin(\beta_2)\right)$$

Let $Z_m = \dfrac{V_m}{I_m} \quad (17)$

Where Z_m is the impedance of the mitigating loop and V_m, volts, is the loop voltage generated by the three-phase current in the loop. The developed approach is well applicable for any voltage rating. After some manipulations and substituting for I_m from

Equation (17), the optimum value of the mitigating loop impedance to create a hundred per cent mitigation of magnetic field at the desired location is expressed by Equation (18)

$$Z_m = \frac{V_m}{I} \left[\frac{\left[\cos(\omega t + \gamma)\right]\left[\dfrac{\cos(\beta_2)}{r_2} + \dfrac{i\sin(\beta_2)}{r_2} - \dfrac{\cos(\beta_1)}{r_1} - \dfrac{i\sin(\beta_1)}{r_1}\right]}{\left(\dfrac{\cos(\omega t)}{R_A}\right)\left(\cos(\alpha_a) + i\sin(\alpha_a)\right) + \left(\dfrac{\cos(\omega t - 120°)}{R_B}\right)\left(\cos(\alpha_b) + i\sin(\alpha_b)\right) + \left(\dfrac{\cos(\omega t + 120°)}{R_C}\right)\left(\cos(\alpha_c) + i\sin(\alpha_c)\right)} \right] \quad (18)$$

Even though, for the numerical demonstration configuration of symmetrically located power lines are utilized, but the developed method is well applicable to the configuration that may lack the symmetry. The established method not only causes a hundred percent cancellation of the magnetic field at any location of desire, but it also causes a simultaneous considerable reduction of the magnetic fields at the other locations.

Since the designed auxiliary mitigating loop is of passive type, no especial feedback equipments are required to compensate for the changes of the load currents, as is the case for active circuit.

Length of the auxiliary mitigating loop can be easily adjusted to suit the spacing between the two adjacent towers. Installing the auxiliary mitigating loop especially beneath the two outer phases of the power line would not create any inconvenience, but due to ice, wind and other possible factors the strength of the tower must be thoroughly checked. Wind gusting at a very high speed may possibly cause a disturbance to the mitigated magnetic field. Corona is also another factor that must be seriously considered when installing this loop. Voltage rating of the capacitor in the loop must be thoroughly investigated to insure the satisfactory operation. Ground clearance is another factor that must be carefully studied.

The developed method is also well applicable to the cases when the auxiliary mitigating loop is installed above the three-phase transmission line. Though such installation would not be associated with the ground clearance, but protection of the auxiliary conductors against the lightning may require serous attention.

The relative effectiveness of the two positions will be discussed in more details in the coming chapters.

The developed method is also well applicable to the situation when the auxiliary mitigating loop is placed at the ground level. Practically, these conductors will be carefully insulated and will be buried at a small depth.

2.5. Numerical Illustrations

In order to illustrate the implementation of the developed Equations as well as the explanations furnished in the past, a flat configuration of a 230 KV transmission line has been utilized, as shown in Figure 3.

The three phases are shown by A, B, and C respectively. $M_1 - M_2$ represent the auxiliary mitigating loop installed beneath the conductors A and C. The ground wires are depicted by $G_1 - G_2$

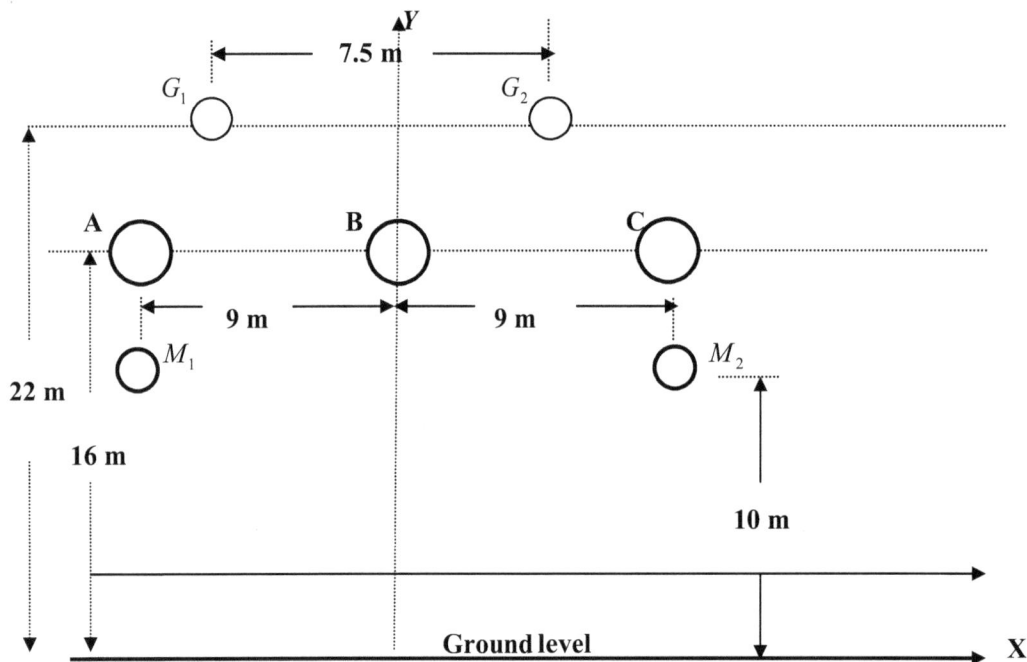

Figure 3. Flat configuration of a 230 KV transmission line.

For the purpose of clarity, magnetic fields are categorized into three.

1- Unmitigated magnetic field is one that is contributed by the transmission line.
2- Mitigating magnetic field is the one generated by the auxiliary mitigating loop.
3- Mitigated magnetic field is resultant of the two magnetic fields.

Point of consideration is the location where hundred percent cancellation of the unmitigated magnetic field is desired.

The total unmitigated magnetic field is the vector sum of the three magnetic fields produced by each phase. Through out this study, the selected locations are assumed to be one meter above the ground, but the developed method is well capable to any point in the space. In order to calculate the resultant unmitigated magnetic field, the unmitigated field produced by each phase of the three-phase transmission line is scrutinized separately.

Let us consider phase A and let the point of consideration be P, having coordinate of $x_i = 9m$ and $y_i = 1m$ as has been depicted in Figure 4.

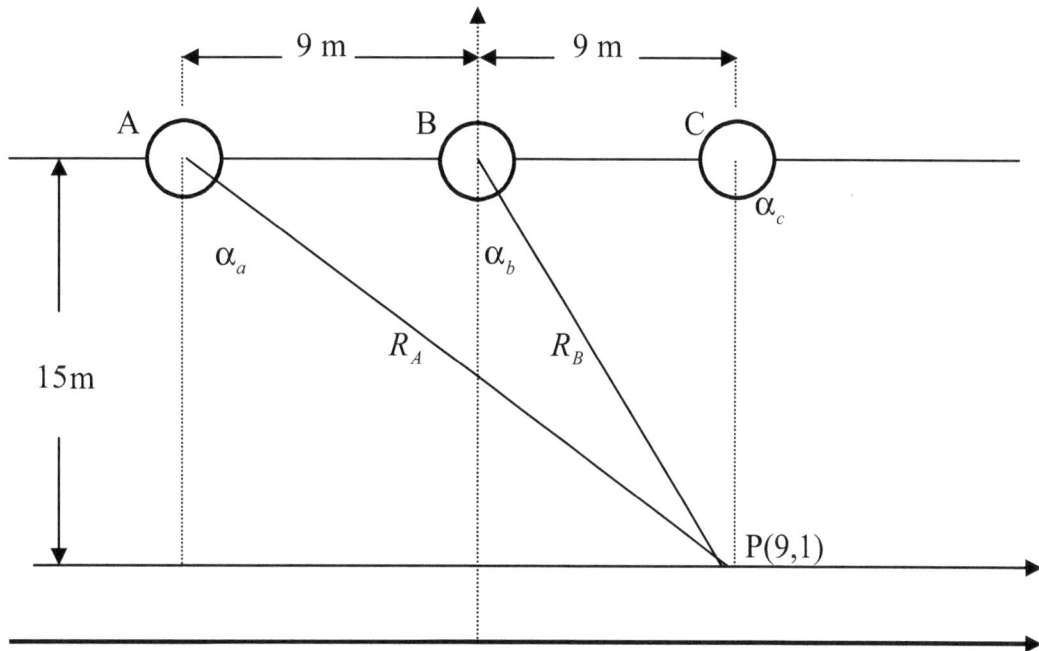

Figure 4. Geometrical location of the three phases with respect to the point of consideration.

In order to be able to determine the unmitigated magnetic field produced by this configuration, first the angular frequency at which maximum field occurs is calculated. Let us calculate all the parameters of Equation (11).

$$R_A = sqrt\left((18)^2 + (15)^2\right)$$

$$R_B = sqrt\left((9)^2 + (15)^2\right)$$

$$R_C = sqrt\left((0)^2 + (15)^2\right)$$

$$\cos(\alpha_a) = \frac{-15}{R_A}$$

$$\sin(\alpha_a) = \frac{-18}{R_A}$$

$$\cos(\alpha_b) = \frac{-15}{R_B}$$

$$\sin(\alpha_b) = \frac{-9}{R_B}$$

$$\cos\left(\alpha_c\right) = \frac{-15}{R_C}$$

$$\sin\left(\alpha_c\right) = 0$$

Substituting in Equation (11), therefore $\omega t = 39.2656$ degrees.

Now implementing Equation (10)

$$\bar{H}_T = 2.4381 - 2.2051\mathrm{i}$$

CHAPTER 3

Fundamental Calculations of Auxiliary Loop Voltage

Abstract: Mitigating magnetic field is caused by the mitigating current in the auxiliary mitigating loop. This current is achieved by dividing the mitigating loop voltage by the mitigating loop impedance. The mitigating loop voltage is the result of the flux induced by each phase of the three phases of the transmission line.

In this chapter, the flux induced by each phase has been thoroughly investigated and the geometrical location of the auxiliary mitigating loop with respect to the power line has been scrutinized and the related equations are established. The vector sum of these three fluxes results in obtaining the total flux penetrating through the mitigating loop and the corresponding equation is set. Finally, an equation to calculate the mitigating loop voltage is developed. It is worth mentioning that phases A, B, and C are at 0°, -120° and +120° respectively.

3.1. Loop Voltage

Loop voltage is the result of flux induced in a loop by a current carrying conductor. In case of a three-phase transmission line, each phase induces its own flux in the loop installed in vicinity of the power line. The vector sum of these three fluxes constitutes the total flux induced by the transmission line.

Let us consider phase A;

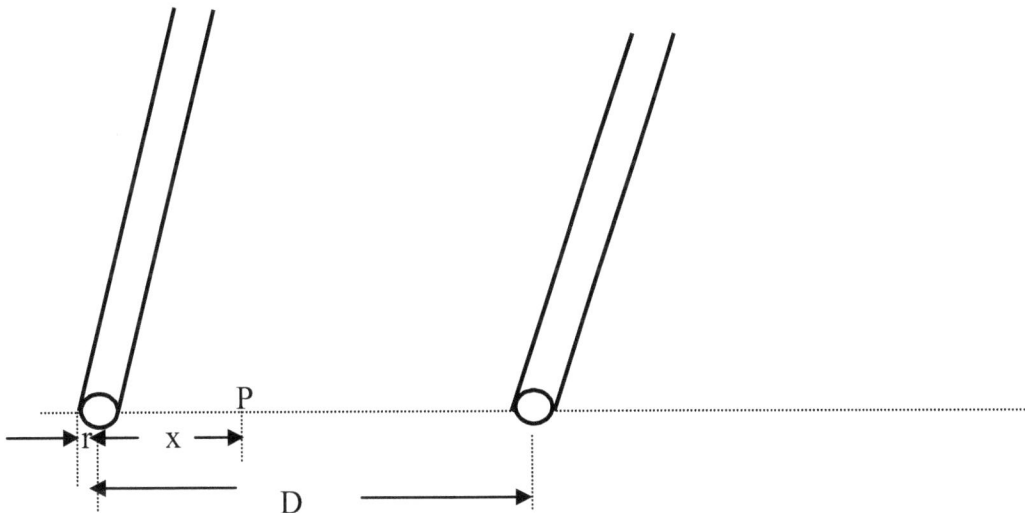

Figure 5. Two conductors forming a loop separated by a distance of D meters.

Figure 5 shows a loop formed by two conductors having radius of r separated by a distance of D. Let P be any arbitrary point placed at a distance of x meters, as shown in the same Figure.

Flux density $\bar{B}$ produced by current, $\bar{I}_A$, of phase A has, at any instance, two components, horizontal and vertical. It is the vertical component that penetrates through the loop.

A.R. Memari (Ed.)

Accordingly;

$$\bar{\Phi}_A = \int_r^D \frac{2*10^{-7}*\bar{I}_A}{x}.dx(19)$$

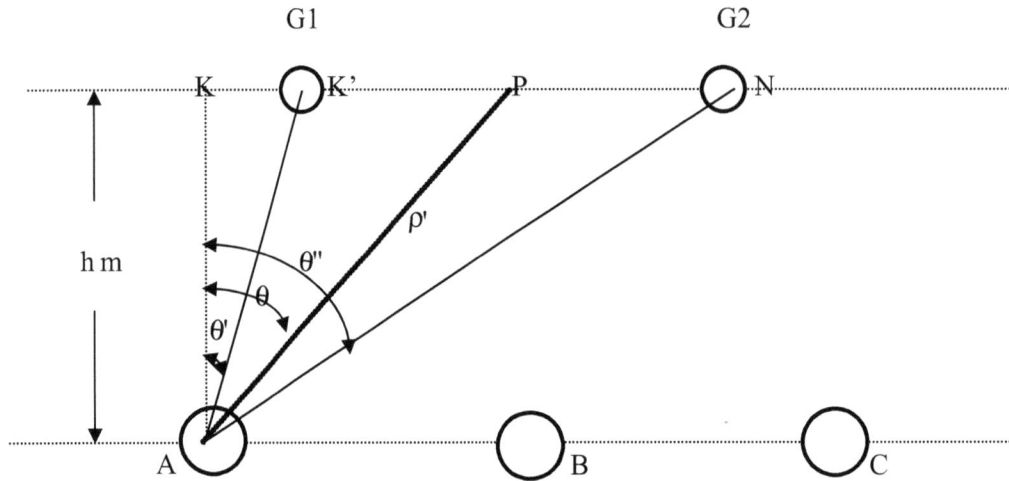

Figure 6. Induced flux in G1-G2 loop, when phase A is considered.

Figure 6 shows a loop made of two parallel conductors placed at an altitude of h meters from a three-phase transmission line. ρ' indicates distance between point P and center of phase A. ρ' makes an angle of θ with the vertical line AK. As this Figure shows, angle θ can vary from θ' to θ'' and x = K'P.

x= KP – KK'

where; $KP = h*\tan(\theta)$

and

$$KK' = h*\tan(\theta')$$

Substituting for the corresponding values in the above Equation;

$$x = h*\tan(\theta) - h*\tan(\theta')$$

as Figure 6 shows, angle θ is the only variable in the above Equation. Therefore, derivative of this Equation with respect to angle θ results in;

$$dx = h*\left(\frac{1}{\cos^2(\theta)}\right)d\theta$$

From the same Figure;

$$h = \rho'* \cos(\theta)$$

Substituting these values in Equation (19);

$$\bar{\Phi}_A = \int_{\theta'}^{\theta'} \left(\frac{2*10^{-7}*\bar{I}_A}{\rho'} \right) (\rho'*\cos(\theta)) \left(\frac{1}{\cos^2(\theta)} \right) d\theta$$

After simplification;

$$\bar{\Phi}_A = 2*10^{-7}*\bar{I}_A * \int_{\theta'}^{\theta''} \frac{\sin(\theta)}{\cos(\theta)}.d\theta$$

Integration of the above Equation results in achieving the flux induced into the loop by the current of phases A, as illustrated below.

$$\bar{\Phi}_A = 2*10^{-7}*\bar{I}_A * [\ln(\cos(\theta')) - \ln(\cos(\theta''))](20)$$

Where:

$$\bar{I}_A = I[\cos(\omega t) + i*\sin(\omega t)]$$

For the purpose of investigating contribution of current $\bar{I}_B$ of phase B to the same loop, let us consider Figure 7.

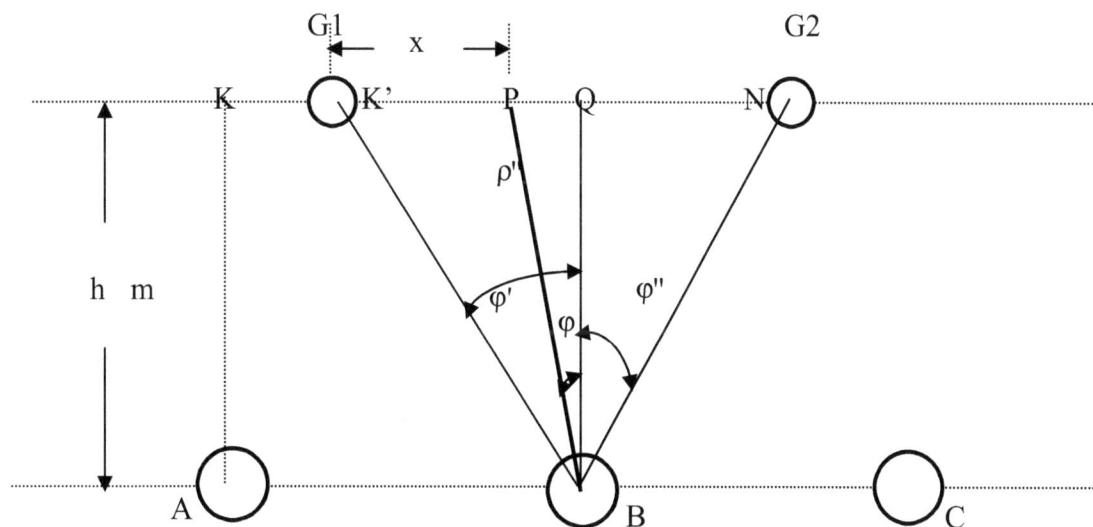

Figure 7. Induced flux in G1-G2 loop, when phase B is considered.

As can be observed, ρ'' the distance from center of phase B to the arbitrary point P makes an angle of φ with the vertical line BQ. This angle is allowed to vary from φ' to φ", as depicted in Figure 7.

The flux induced by $\bar{I}_B$ in the loop is given by Equation (21)

$$\bar{\Phi}_B = \int_r^D \frac{2*10^{-7}*\bar{I}_B}{x}.dx (21)$$

From this Figure;

x = K'Q – PQ

where $K'Q = h * \tan(\phi')$

and

$$PQ = h * \tan(\varphi)$$

Therefore:

$$x = h * \tan(\varphi') - h * \tan(\varphi)$$

In the above Equation φ is the only variable, consequently;

$$dx = -h * \left(\frac{1}{\cos^2(\varphi)}\right) d\varphi$$

$$h = \rho'' * \cos(\varphi)$$

Substituting all the above relationships in Equation (21), after some manipulations

$$\bar{\Phi}_B = 2*10^{-7}*\bar{I}_B*(\ln(\cos(\varphi')) - \ln(\cos(\varphi''))) (22)$$

Where

$$\bar{I}_B = I[\cos(-120°) + i * \sin(-120°)]$$

The flux contributed by phase C to the same loop is depicted by Equation (23)

$$\bar{\Phi}_C = \int_r^D \frac{2*10^{-7}*\bar{I}_C}{x}.dx (23)$$

In order to find the contribution of $\bar{I}_C$ in inducing flux in the G1-G2 loop, similar procedures are followed and consequently;

$$\bar{\Phi}_C = 2*10^{-7}*\bar{I}_C*(\ln(\cos(\Psi')) - \ln(\cos(\Psi''))) (24)$$

$$\bar{\Phi}_C = 2*10^{-7}*\bar{I}_C*\left(\ln\left(\cos\left(\Psi'\right)\right)-\ln\left(\cos\left(\Psi''\right)\right)\right)(24)$$

Where;

$$\bar{I}_C = I\left[\cos\left(120°\right)+i*\sin\left(120°\right)\right]$$

The resultant flux induced by this transmission line is the vector sum of the three fluxes.

$$\bar{\Phi}_T = \bar{\Phi}_A + \bar{\Phi}_B + \bar{\Phi}_C\left(25\right)$$

Consequently, the loop voltage is given by Equation (26)

$$\bar{V}\_loop = j\omega\bar{\Phi}_T\ell\left(26\right)$$

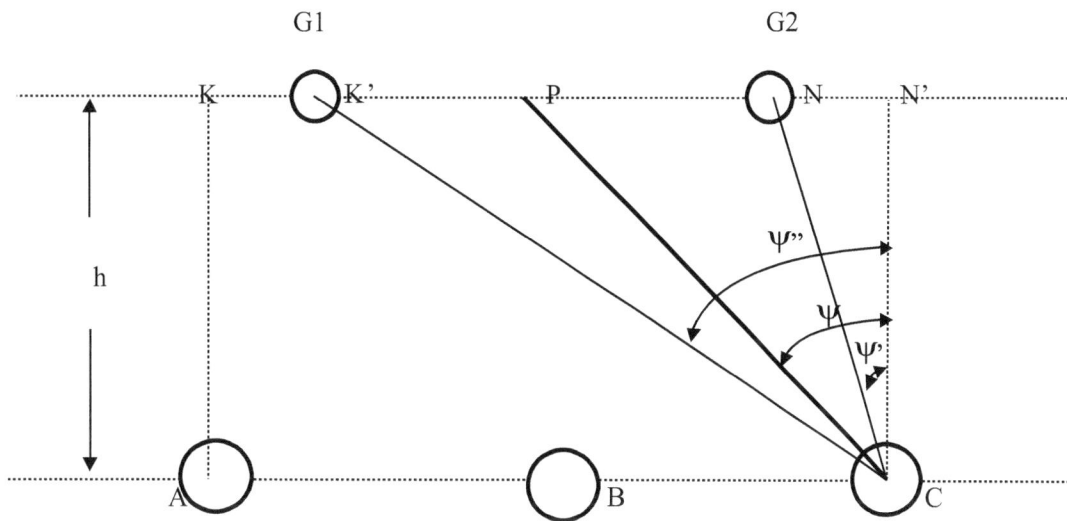

Figure 8. Geometrical position of phase C to the same loop G1-G2.

CHAPTER 4

Calculation of Mitigating Magnetic Field

Abstract: A numerical illustration has been set up to demonstrate the applicability of the developed equations to calculate the mitigating magnetic field. A point in the space having a coordinate of (9, 1) has been selected as the point of consideration.

The total flux induced by the three phases of the transmission line is calculated, from which the mitigating loop voltage is achieved. In order to achieve our attempt of establishing a hundred percent cancellation of the magnetic field, optimal value of the loop impedance is determined.

Variation of unmitigated magnetic field over one complete cycle in depicted and simultaneous variation of unmitigated magnetic field and mitigated magnetic field is also shown.

The influence of the mitigating magnetic field produced by the loop on the other locations is also investigated and the results are tabulated. Effect of angular frequency on producing magnetic field at other location is studied.

4.1. Mitigating Magnetic Field

Mitigating magnetic field is one, which is produced by an auxiliary loop located either below or above the outer phases of a three-phase transmission line. Figure 9 shows the case when the mitigating auxiliary loop is located beneath the line. The already existing ground wires shown by G_1 and G_2, in the Figures 6 through 8, could be used to also act as a mitigating loop. Effectiveness of such a loop will be discussed in full details in the coming chapter.

As was explained in the previous chapter, in order to achieve a hundred percent cancellation of the unmitigated magnetic field, the magnetic field produced by the auxiliary mitigating loop must be equal in magnitude to that of unmitigated field, but opposite in direction. The auxiliary loop used for this purpose is of passive type, therefore the flux induced by the three-phase currents is fully responsible to generate the mitigating current in the mitigating loop. This is the advantage of having a passive loop, because with changes in the load current not only the magnetic fields produced by the line is affected, but also the mitigating current. Consequently, the resultant mitigated magnetic field remains unchanged.

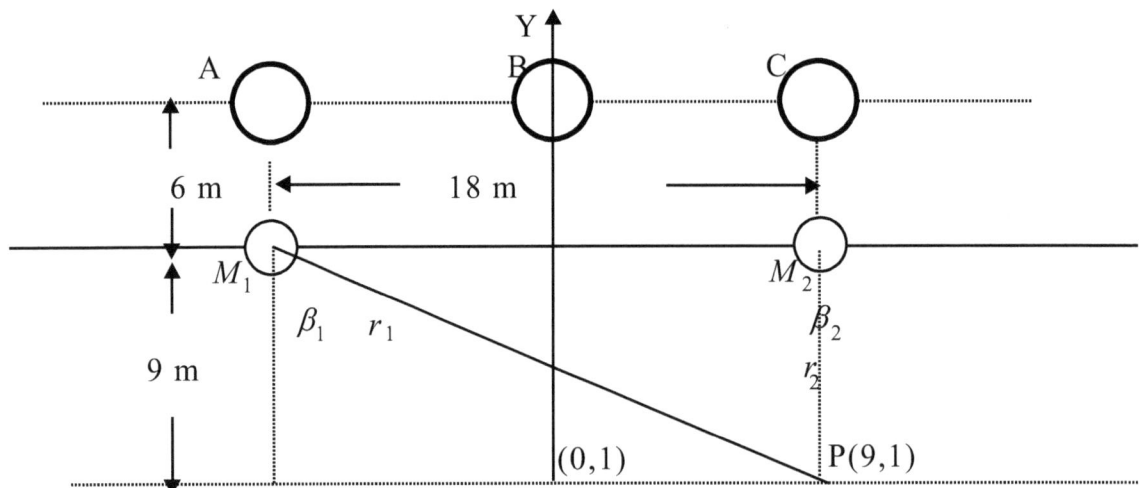

Figure 9. Geometrical location of mitigating loop with respect to point P.

For the illustrative purpose, a specific case has been analyzed. The mitigating loop is located at 6 meters beneath the power line, as shown in Figure 9. Point P is the point of consideration located at (9,1) from the center of right-of-way.

$$r_1 = \sqrt{(9)^2 + (18)^2} = 20.1246$$

$$r_2 = \sqrt{(9)^2 + (0)^2} = 9$$

$$\cos(\beta_1) = \frac{-9}{20.1246}$$

$$\sin(\beta_1) = \frac{-18}{20.1246}$$

$$\cos(\beta_2) = -1$$

$$\sin(\beta_2) = 0$$

Implementation of Equation (16) requires us to determine the value of I_m, which means that the optimal value of the mitigating loop impedance Z_m of Equation (18) must first be obtained. All the parameters of Equation (18), but except V_m, the mitigating loop voltage, have already been determined. In order to calculate V_m, let us apply Equations (20), (22), (24), (25) and (26).

As shown in Figure 10;

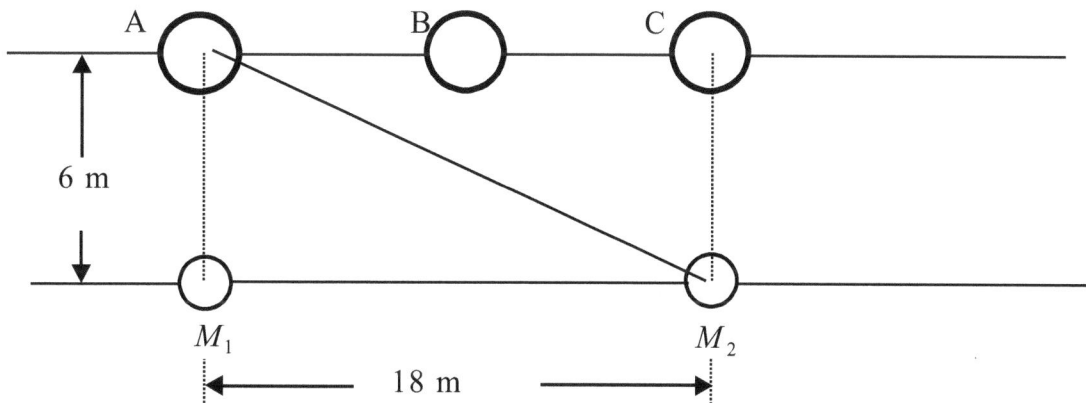

Figure 10. Geometrical location of mitigating loop conductors with respect to the three phases.

Figure 10;

$$\theta_a' = 0$$

$$\theta_a'' = 1.2490$$

$$\theta'_b = 0.9828$$

$$\theta''_b = 0.9828$$

$$\theta'_c = 1.2490$$

$$\theta''_c = 0$$

$$I_A = 460\,Amps$$

$$I_B = 460*\left[\cos(-120°) + i*\sin(-120°)\right]$$

$$I_C = 460*\left[\cos(120°) + i*\sin(120°)\right]$$

$$\bar{\Phi}_A = 2*10^{-7}*\bar{I}_A*\left[\ln(\cos(\theta'_a)) - \ln(\cos(\theta''_a))\right]$$

$$\bar{\Phi}_B = 2*10^{-7}*\bar{I}_B*\left[\ln(\cos(\theta'_b)) - \ln(\cos(\theta''_b))\right]$$

$$\bar{\Phi}_C = 2*10^{-7}*\bar{I}_C*\left[\ln(\cos(\theta'_c)) - \ln(\cos(\theta''_c))\right]$$

The total flux penetrated by the three phases A, B and C would be the vector sum of the three fluxes. Therefore;

$$\bar{\Phi}_T = \bar{\Phi}_A + \bar{\Phi}_B + \bar{\Phi}_C$$

and

$$V_m = 2000*\pi*60*\bar{\Phi}_T$$

Therefore, for the considered case, the mitigating loop voltage would be equal to 69.1616 volts.

Since the main purpose of our attempt is to establish hundred percent cancellation of the magnetic field, optimal value of the loop impedance, which is responsible to generate the mitigating current in the mitigating loop, must be determined. Applying Equation (18) and allowing angle $(\gamma) = -(\omega t)$, the value of the impedance for the given case would be equal to $\bar{Z}_m = -0.3207 - 0.0889i$. Implementation of Equation (16) sets the mitigating magnetic field equal to $\bar{H}_m = -2.4406 + 2.2023i$.

The vector sum of unmitigated and mitigating magnetic fields results in the mitigated magnetic field.

Variation of unmitigated magnetic field over one complete cycle is depicted in Figure 11.

This Figure shows that mitigating magnetic field forms the major axis of the unmitigated field ellipse, which indeed justifies the previous explanations. As this Figure shows, magnitude of mitigating magnetic field is equal to that of unmitigated magnetic field with their orientations in opposite directions, resulting in a zero mitigated magnetic field. The minor axis of the unmitigated magnetic field ellipse is nothing but the mitigated magnetic field.

Simultaneous variations of unmitigated magnetic field and mitigated magnetic field versus angular frequency ωt in degrees are shown in Figure 12. As this Figure shows, the mitigated magnetic field achieves zero values at angles 39.3 degrees and 219.3 degrees respectively. These are the same angles at which unmitigated magnetic field obtains its maximum value of 3.3 A / m. This Figure also shows that at the angular frequencies which unmitigated magnetic field achieves its minimum values, the mitigated magnetic field obtains its maximum values of 0.82 A/m.

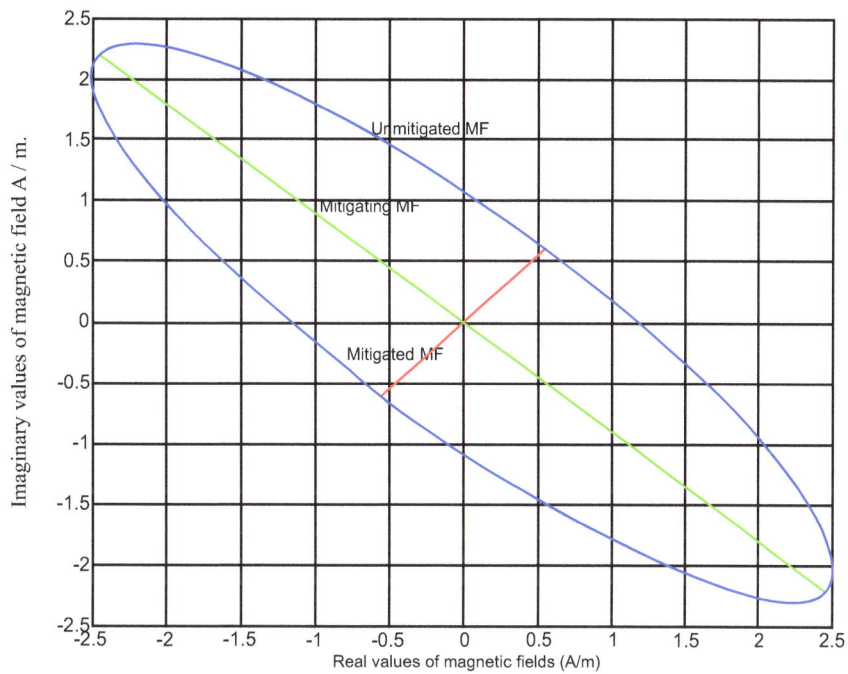

Figure 11. Vector positions of the three magnetic fields for x=9 m,y =1 m.

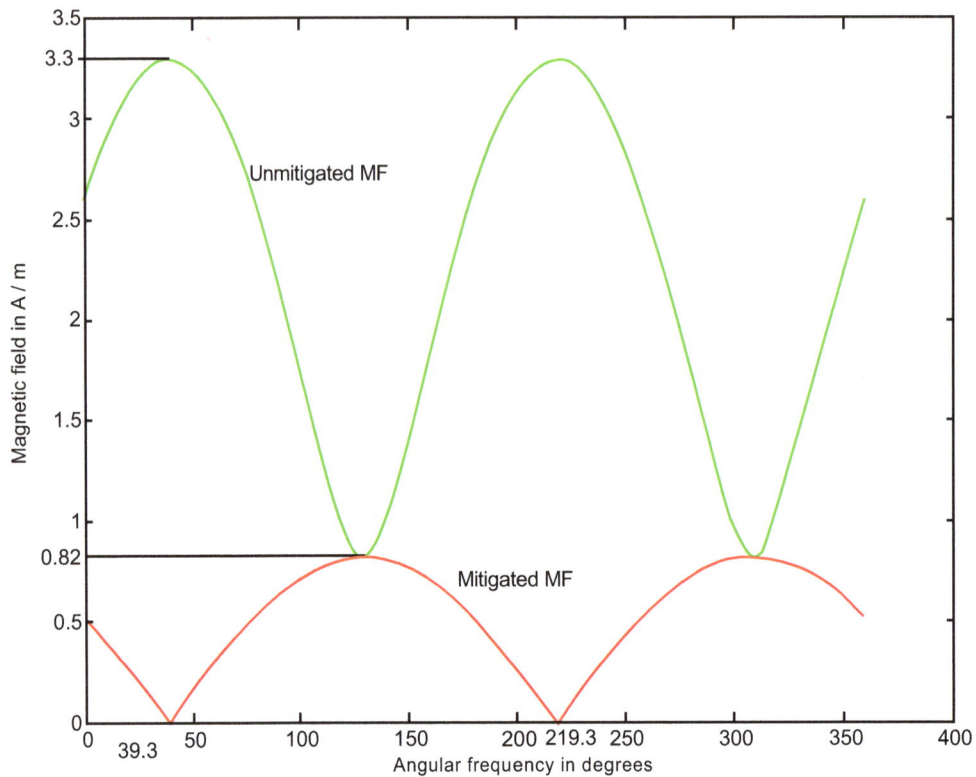

Figure 12. Variations of magnetic fields with respect to angular frequency.

4.2. Mitigations at other Locations

The mitigating magnetic field produced by the auxiliary loop influences the magnetic fields produced by the three-phase line at other locations. In order to demonstrate contribution of this field on magnetic fields generated by the lines, four different points such as $P_1(0,1)$, $P_2(36,1)$, $P_4(-36,1)$, $P_5(-9,1)$ are taken under consideration. Since the aim is to scrutinize the effect of mitigating magnetic field, which is generated to produce hundred percent cancellation of magnetic field at point $x_j = 9$ m and $y_j = 1$ m, the angular frequency at which magnetic field at this point achieves its maximum value is utilized to calculate the magnetic fields at the other four locations.

For further clarity of this explanation, let us investigate location (36,1).

ωt = 39.2656 degrees.

$$R_A = sqrt\left((45)^2 + (15)^2\right) = 47.4342 \text{ meters}$$

Similarly, R_B = 41.7852 m and R_C = 30.8869 m

$\cos(\alpha_a) = -15/47.4342$ and $\sin(\alpha_a)$ = -45/47.4342. Applying the geometrical location of the point P with respect to the three phases, other parameters of Equation (10) are calculated.

Implementation of Equation (10) results in unmitigated magnetic field of 0.5824 + 0.5252i

Now, apply Equation (16) and calculate mitigating magnetic field, which would be equal to -0.3241 - 0.3206i. The mitigated magnetic field, which is vector sum of these two fields would be equal to 0.33 A / m, as shown in Figure 13.

The vector sum of this mitigating magnetic field and unmitigated magnetic field produced at the other locations results in simultaneously obtaining the mitigated magnetic field at each point. As Figure 13 shows, in addition to establishing a-hundred percent mitigation at point of consideration (9,1), noticeable amount of mitigations have been obtained at the other four locations as also shown in Table 1.

Figure 14 and Table 1 reveal that the unmitigated magnetic fields produced at the left hand side of the center of the right – of – way are slightly less than that of symmetrically allocated points on the right hand side of the center of the right- of – way.

4.3. Effect of Angular Frequency

In the process followed, the angular frequency of value 39.3° has been used to calculate the value of unmitigated magnetic field at the other three points. This is the angular frequency responsible to create maximum unmitigated field at $x_j = 9m$, $y_j = 1m$

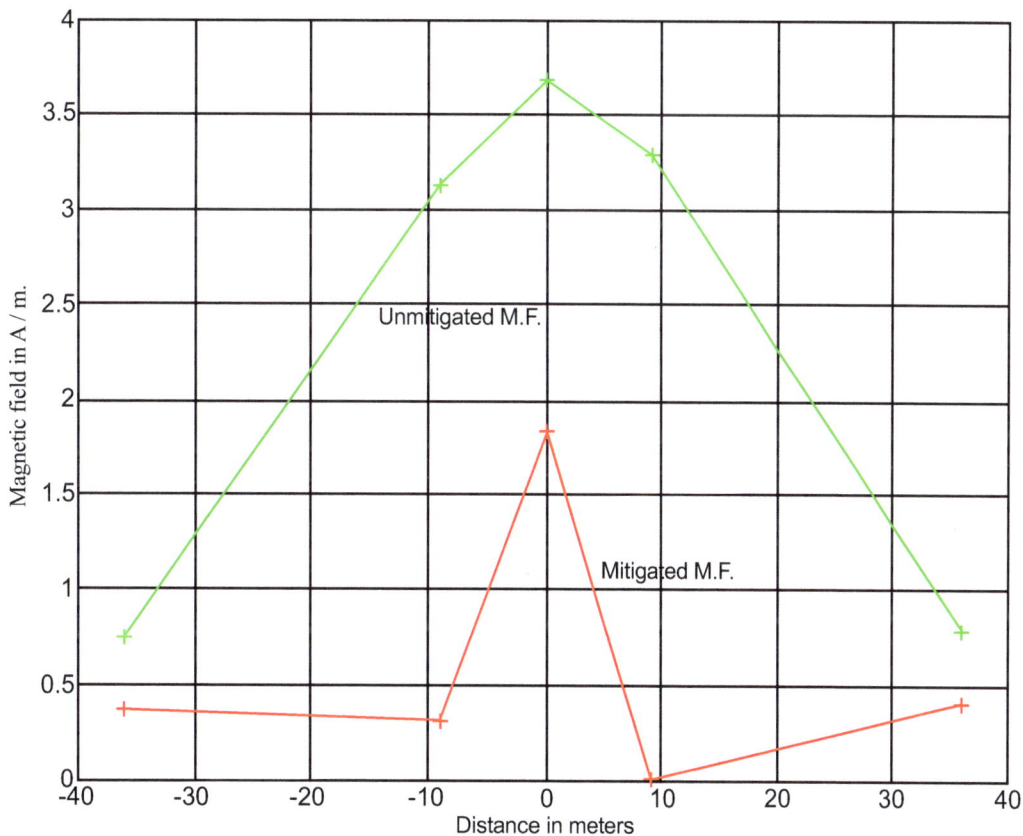

Figure 13. Contribution of mitigating M.F. to the other points.

Table 1 shows the numerical values for unmitigated and mitigated magnetic fields at five different locations. According to this Table, 90 percent of mitigation has been achieved at symmetrically located point of x_j = -9 m and y_j = 1 m, with point x_j = +9 m and y_j = 1 m as the point of consideration.

As this Table shows, the symmetrically located points such as (9,1) and (-9,1), (36,1) and (-36,1) do not possess the same value of unmitigated magnetic field. The second row of Table 3 and Figure 14 show the same values for symmetrically located points. These values of unmitigated magnetic fields are obtained with ωt of Equation (10) equal to the value responsible to produce maximum value of unmitigated magnetic field, as shown in Table 2.

TABLE **1.** Corresponding unmitigated and mitigated M.F. at five different locations

Distances from center of right-of-way in meters.	-36	-9	0	9	36
Unmitigated M.F. (A/m)	0.7534	3.1277	3.6868	3.2874	0.7842
Mitigated M.F.	0.3721	0.3106	1.8436	0.000	0.4008
Percentage of mitigation	50.606	90.0701	49.9941	100	48.8957

The amount of generated magnetic field by the three-phase line drastically drops as one moves further away from the center of the right-of-way.

TABLE **2.** Angular frequency at which maximum unmitigated magnetic fields occurs at the corresponding location

Distances from center of right-of-way, meters.	-45	-36	-27	-18	-9	0	9	18	27	36	45
ωt in degrees	24	23	21.5	20.2	20.7	30	39.3	39.9	38.4	37	36

CHAPTER 5

Geometrical Positions of Auxiliary Loop

Abstract: Effect of geometrical location of the auxiliary mitigating loop with respect to the three phases of the transmission line is scrutinized. In this process, the mitigating loop is first placed above and then below the two outer phases of the power line and a comparative approach has been established. Positions of maximum and minimum values of the mitigated magnetic fields with respect to the geometrical location of the loop are also investigated. A 230 KV flat transmission line has been utilized and effect of mitigation at the other points has also been studied.

Process of mitigation with respect to the geometrical location of the auxiliary loop and the correlation between the three types of magnetic fields has also been studied and the related figures and Tables are depicted.

5.1. Geometrical Locations and Magnetic Field

There are two elements that play very important role in producing magnetic field, current and distance. In the case of mitigating magnetic field, in addition to these two elements the altitude of the auxiliary mitigating loop from the three-phase transmission line and also the separation between the two conductors shaping the auxiliary mitigating loop are also immensely effective and should be thoroughly studied.

In order to illustrate effectiveness of geometrical location of the auxiliary mitigating loop on mitigated magnetic field, two different locations, above and below the transmission line, are selected and then at each location the mitigating loop is allowed to change its altitude with respect to the power line.

TABLE 3. Effect of variations of mitigating loop altitude with respect to the power lines.

Distance from center of right-of-way. Meters		-45	-36	-9	0	9	36	45
Unmitigated M.F. A/m		0.5249	0.7848	3.2874	3.7296	3.2874	0.7848	0.5249
Mitigated Magnetic Fields. A/m	h = 20 m	0.0350	0.0685	0.4158	0.000	0.4158	0.0685	0.0350
	h = 10 m	0.0227	0.0468	0.2952	0.000	0.2952	0.0468	0.0227
	h = 6 m	0.0165	0.0202	0.2092	0.000	0.2092	0.0355	0.0165
	h =5 m	0.0149	0.0322	0.1818	0.000	0.1818	0.0322	0.0149
	h =4m	0.0132	0.0287	0.1515	0.000	0.1515	0.0287	0.0132
	h =3m	0.0114	0.0247	0.1181	0.000	0.1181	0.0247	0.0114
	h =2m	0.0093	0.0197	0.0816	0.000	0.0816	0.0197	0.0093
	h =1m	0.0060	0.0119	0.0425	0.000	0.0425	0.0119	0.0060

Figure 15 shows a schematic diagram of a 230 KV transmission line with its phases separated by 9 meters. The mitigating loop is installed at 6 meters above the two outer phases A and C. Let (0,1) be the point of consideration. As Table 2 shows, the angular frequency responsible to generate maximum unmitigated magnetic field at (0,1) is equal to 30 degrees.

Equation (10) is well applicable to determine the total unmitigated magnetic field produced by this system at (0,1). From Figure 15;

$$R_A = sqrt\left((9)^2 + (15)^2\right) = 17.4929m$$
$$R_B = 15m$$
$$R_C = R_A$$

$$\cos(\alpha_a) = \frac{-15}{R_A}$$
$$\sin(\alpha_a) = \frac{-9}{R_A}$$
$$\cos(\alpha_b) = \frac{-15}{15} = -1$$
$$\sin(\alpha_b) = 0$$
$$\cos(\alpha_c) = \frac{-15}{R_C}$$
$$\sin(\alpha_c) = \frac{9}{R_C}$$

I = 450 Amps.

Therefore

$$\vec{H}_T = -0.0000 - 3.6485i$$

Implementation of Equation (16) results in obtaining value of the mitigating magnetic field.

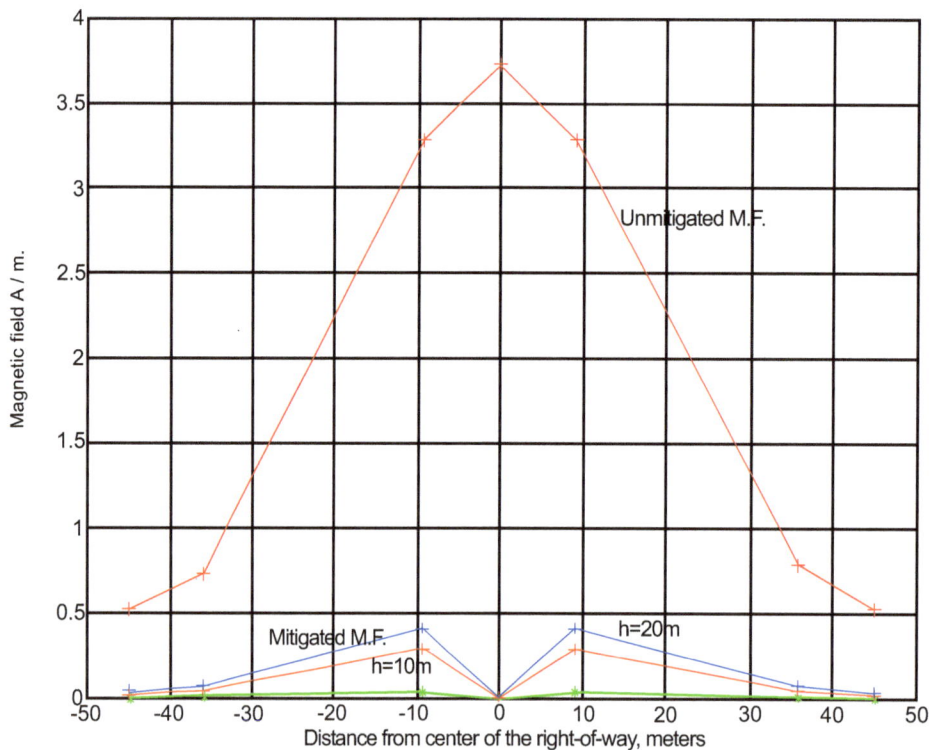

Figure 14. Contribution of altitude variation on mitigated magnetic field.

Case one;

5.2. The Auxiliary Mitigating Magnetic Field is Located at Above the Power Lines.

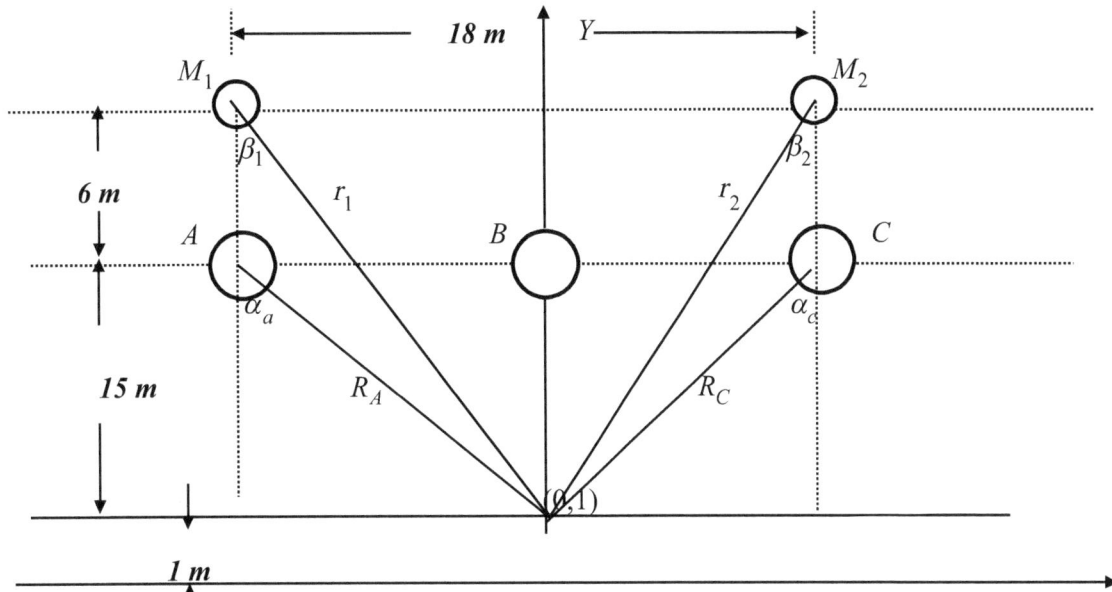

Figure 15. Schematic diagram of transmission line with mitigating loop $M_1 - M_2$.

From Figure 15;

$$r_1 = sqrt\left((9)^2 + (21)^2\right)$$

$$r_2 = r_1$$

$$\cos(\beta_1) = \frac{-21}{r_1}$$

$$\sin(\beta_1) = \frac{-9}{r_1}$$

$$\cos(\beta_2) = \frac{-21}{r_2}$$

$$\sin(\beta_2) = \frac{9}{r_2}$$

V_loop = 69.1616 volts

Since, our aim is to create hundred percent cancellation of the unmitigated field, $\gamma = -\omega t$.

Implementation of Equation (18) results in achieving the optimal value of the loop impedance Z_m, from which I_m is calculated. Substituting the above obtained values in Equation (16)

$$\bar{H}_m = 0.0000 + 3.6485i$$

with

$$\bar{H}_T = -0.0000 - 3.6485i$$

It can be clearly observed that $\bar{H}_m$ and $\bar{H}_T$ possess the same values but their orientations are in the opposite directions. Mitigated magnetic field, which is the vector sum of $\bar{H}_m$ and $\bar{H}_T$, is equal to zero. In order to establish the locus of this unmitigated magnetic field, let ωt vary over one complete cycle of 360°

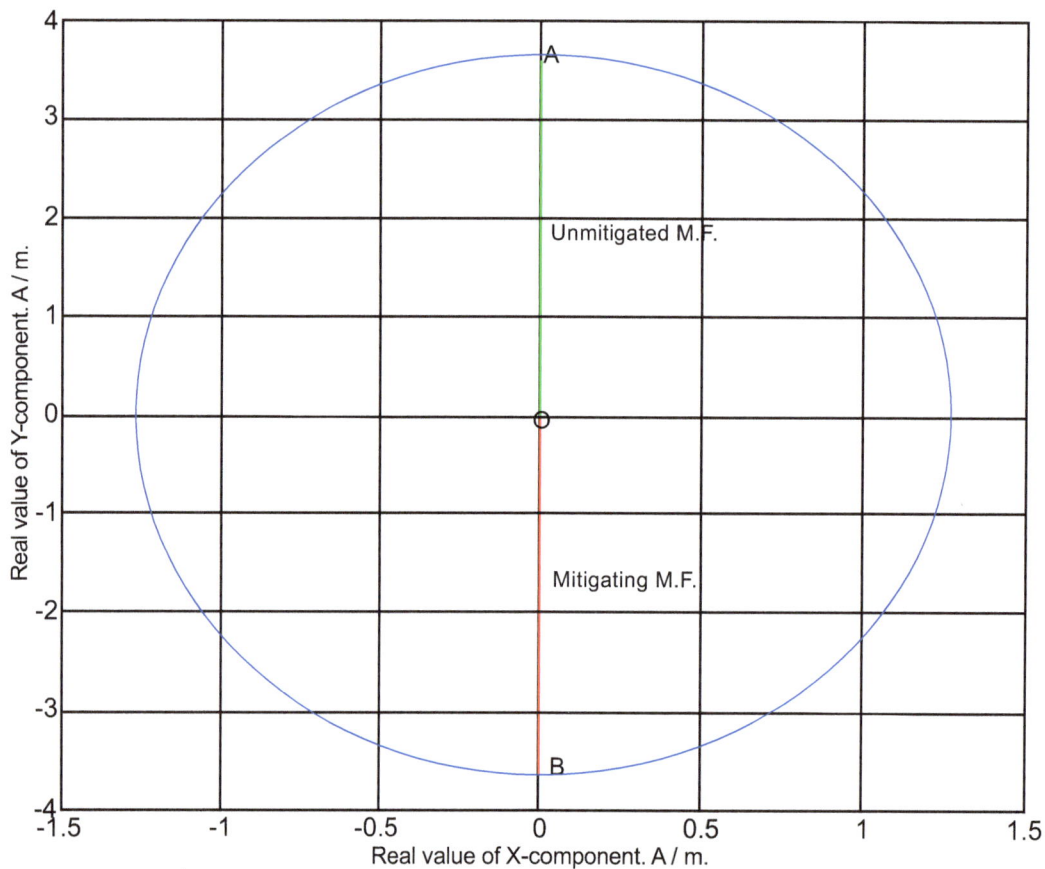

Figure 16. Position of magnetic fields for x=0, y=1 m.

Figure 16 shows position of unmitigated magnetic field OA, which is equal in magnitude to OB, but it is in the opposite direction. As a result, the resultant is equal to zero.

In order to scrutinize the correlation between the mitigated magnetic field and the selected location, let us choose seven points such as –45, -36, -9, 0, 9, 36, and 45 meters from the center of the right-of-way, with $x_j = 0m$, $y_j = 1m$ as the point of consideration.

Let also the mitigating loop change its altitude from one meter above the power line to six meters, with one-meter interval. For further illustration of this approach, the loop is also placed at farther altitude, 10 meters and 20 meters as shown in the Table 3.

The calculations show that the orthogonal distances increase as the altitude is increased, but the loop voltage drops from 173.7260 volts at h=1 m to 69.1616 volts when h= 6 m. Mitigating current responsible to create mitigation at the point of consideration (0,1) is utilized to produce the mitigating magnetic field at the other points. Vector sum of the unmitigated magnetic field and mitigating magnetic field results in mitigated magnetic field at each point.

As the Table 3 shows, values of the mitigated magnetic fields reduce as values of h reduce. This is mainly due to increase of the induced voltage in the auxiliary mitigating loop as this loop gets closer to the transmission lines.

Irrelevant to the position of the auxiliary mitigating loop from the power lines, the mitigated magnetic field at the point of consideration would always be zero, as shown in the same Table.

Figure 17 illustrates values of mitigated magnetic field for seven different locations when mitigating loop selects two different altitudes of h = 10 m and h = 20 m. As this Figure shows, values of mitigated magnetic fields with h = 20 m are greater than that of h =10 m but in both the cases, the value of mitigated magnetic field at the point of consideration (0,1) is always zero.

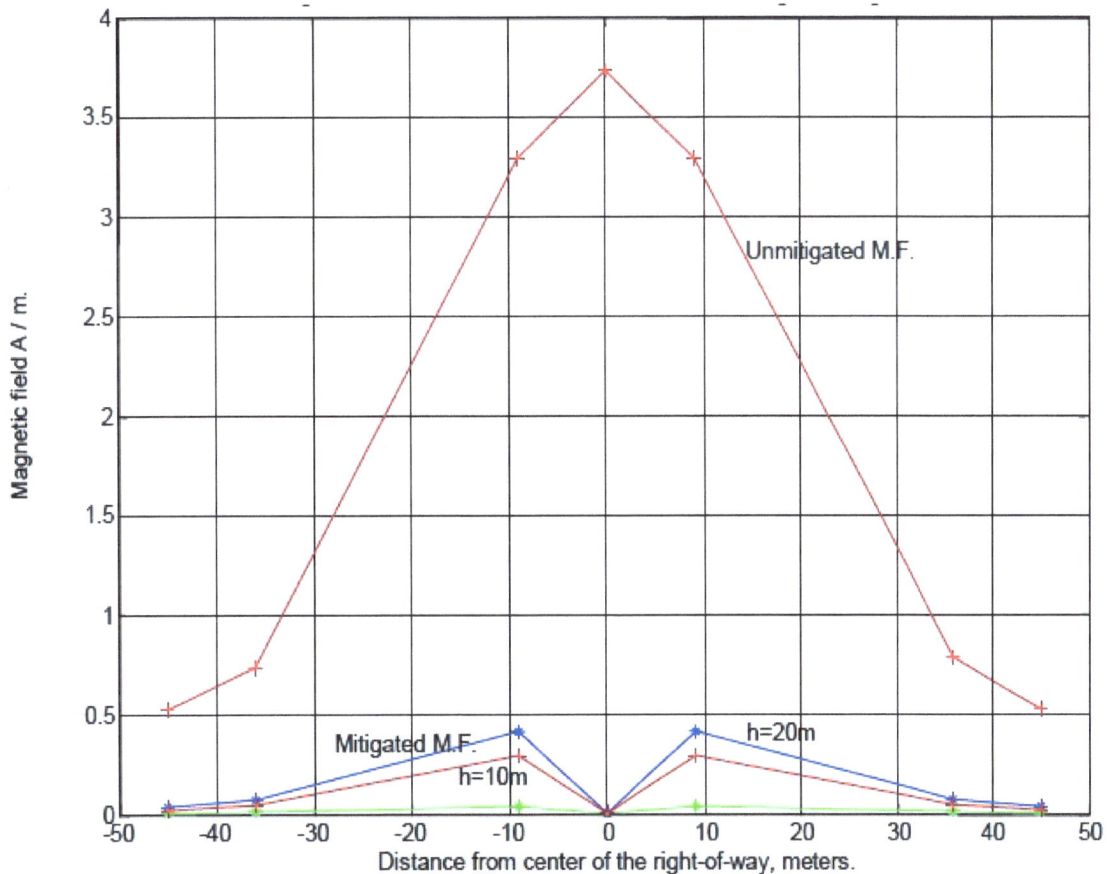

Figure 17. Contribution of altitude variation on mitigated magnetic field.

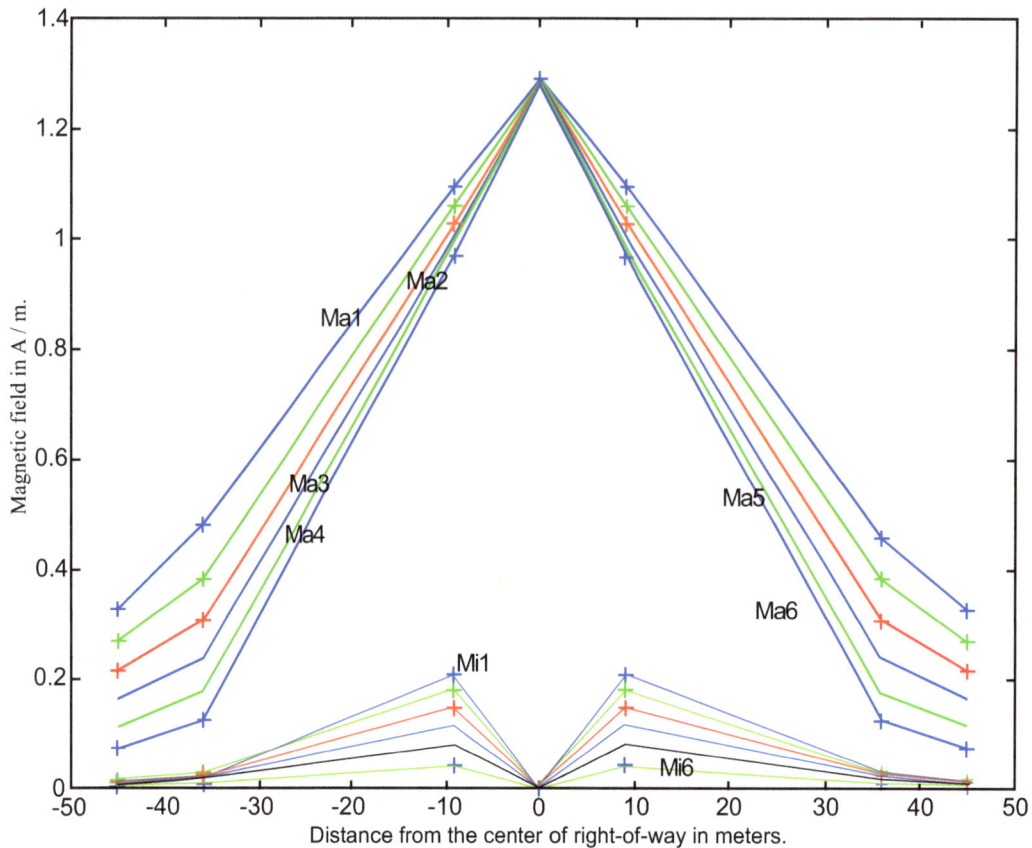

Figure 18. Position of max. min. values of mitigated M.F. as h is varied.

5.3. Effect of Altitude on the Fields

Figure 18 shows relative positions of minimum and maximum values of the mitigated magnetic fields as altitude, h, is allowed to vary from six meters to one meter above the power lines.

As was already mentioned, the load currents have sinusoidal variation and therefore, so do mitigating currents, as they are the result of induced flux in the auxiliary mitigating loop produced by these three-phase load currents. Mitigated magnetic field, which is the vector sum of total unmitigated and mitigating magnetic fields, has also a sinusoidal variation during one full cycle of 360°. It therefore, causes a hundred percent cancellation of the total unmitigated magnetic field only at a particular angular frequency at which the maximum unmitigated magnetic field is achievable. Variation of angular frequency, ωt, over 360° cause the mitigated magnetic field to obtain sinusoidal variation. At any location, the mitigated magnetic field has a minimum value and a maximum value. These two values correspond to two different values of angular frequency. The difference between these two frequencies would always be 90°, as shown in Table 4. This Table shows maximum and minimum values of mitigated magnetic fields at a specified location of $x_j = 45$ m and $y_j = 1$ m from the center of the right-of-way. This Table also shows that as altitude of the mitigating loop from the two outer phases of the transmission line decreases, minimum and maximum values of the mitigated field also decreases, but difference between ωt always remains at 90°.

Considering the point of consideration (0,1), mitigated magnetic field reaches its zero value only at the angular frequency at which the unmitigated magnetic field obtains its maximum value. Since mitigated magnetic field has a sinusoidal variation, it obtains non-zero values at all the other angular frequencies.

Figure 18 shows minimum and maximum values of mitigated magnetic fields for seven different locations-45, -36, -9, 0, 9, 36, and 45 as the mitigating auxiliary loop's altitude from the power line varies from 6 meters to 1 meter. This Figure shows that at the point of consideration, $x_j = 0$ m and $y_j = 1$ m, mitigated magnetic field achieves zero value irrelevant to altitude h from the power line.

At the point of consideration, mitigating magnetic field achieves a maximum value equal to that of unmitigated magnetic field. Since their orientations are in the opposite directions, their vector sum results in zero value.

Mitigating magnetic field at the point of consideration possesses a zero value. Consequently, the maximum value of mitigated magnetic field is nothing but the minimum value of unmitigated magnetic field. for the given case, this value is equal to 1.2920 A /m. This value is irrelevant to the altitude of the auxiliary mitigating magnetic field from power line.

In this Figure, M_{i_1} to M_{i_6} indicate the minimum values of the mitigated magnetic fields. M_{a_1} to M_{a_6} indicate the maximum values of the mitigated magnetic fields for six different altitudes of the auxiliary mitigating loop.

TABLE **4**. Max. and Min. values of mitigated M.F. and their corresponding angular frequencies at x=45 and y=1 above the power lines.

Distance from the power lines	Minimum Mitigated M.F.		Maximum Mitigated M.F.	
	Values A/m	Angles, degrees	Values A/m	Angles, degrees
h=6 m	0.0065	110	0.3263	20
h=5 m	0.0049	108	0.2704	18
h= 4 m	0.0032	105	0.2160	15
h = 3 m	0.0014	100	0.1640	10
h =2 m	0.0093	91	0.1156	1
h = 1 m	0.0060	71	0.0754	161

5.4 Effect of Point of Consideration

In order to further illuminate effect of position of the mitigating loop as well as selection of point of consideration on mitigated magnetic field, let us select $x_j = 9$ m, $y_j = 1$ m as a new point of consideration.

As Figure 19 shows, when the auxiliary mitigating loop changes its altitude with respect to the power lines installed below the loop, values of the maximum mitigated magnetic fields at point (0,1) change their values from 1.43 A / m to 1.55 A /m.

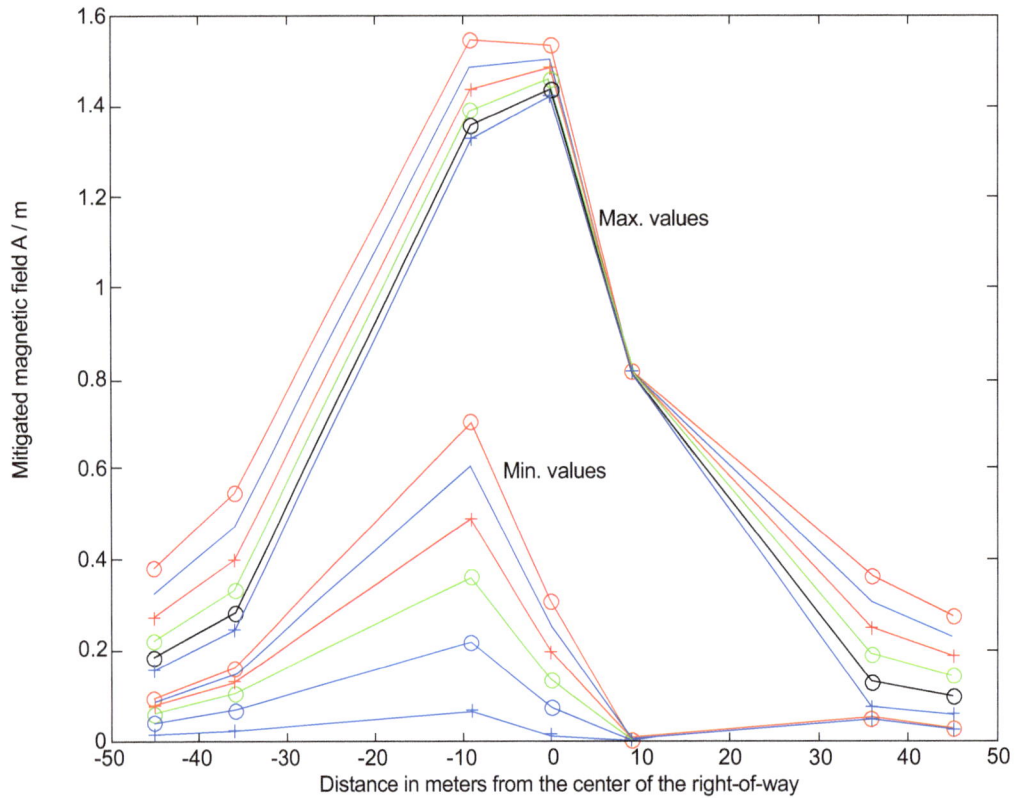

Figure 19. Characteristics of max. and min. values of mitigated M.F., h varies above the line.

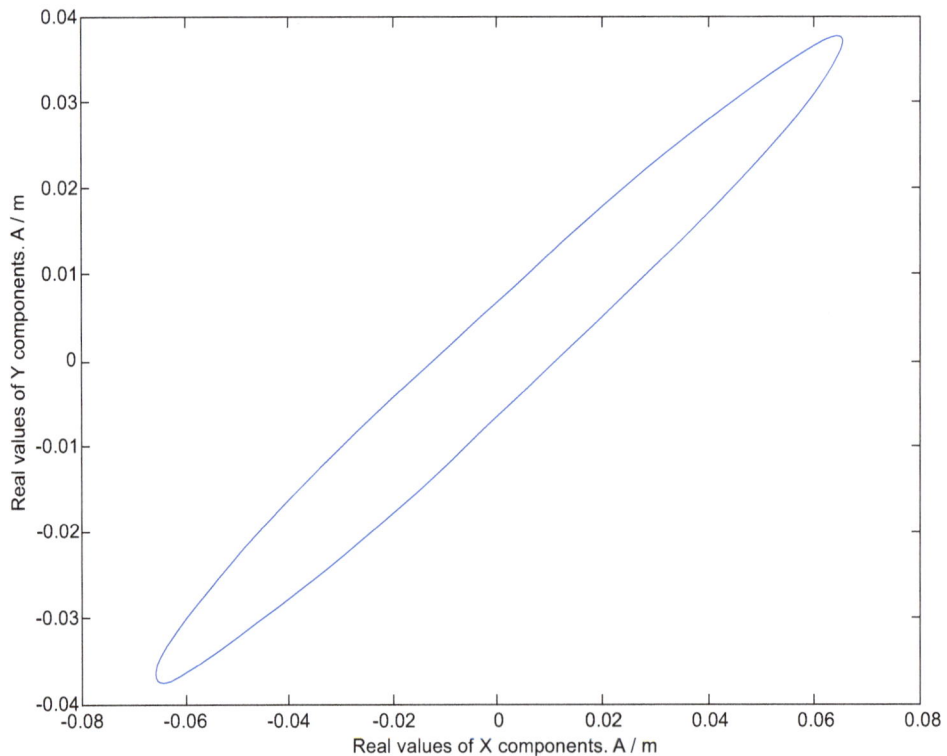

Figure 20. Elliptical trajectory of mitigated M.F. at a specified location (45,1).

Figure 19 also shows that at the point of consideration, $x_j = 9$ m , $y_j = 1$ m, values of maximum mitigated magnetic fields remain constant at 0.8170 A / m. This value is irrelevant to the altitude of the auxiliary mitigating loop from the power line.

Comparing this Figure with Figure 18, it could be concluded that at the point of consideration values of maximum mitigated magnetic fields always remain at a constant value. The reason should be investigated in the fact that at the point of consideration, irrelevant to altitude of the mitigating loop from the transmission lines, the maximum and minimum values of unmitigated and mitigating magnetic field remain unchanged.

As it was previously explained, the mitigated magnetic field has also sinusoidal variation over one complete cycle of 360° and it also possesses minimum value as well as maximum value. Such a variation is depicted in Figure 20 for a specific location of $x_j = $ 45 m and y_j = 1 m. for the case when the mitigating loop is installed above the transmission lines. This Figure shows elliptical trajectory of the mitigated magnetic field.

Unlike mitigated magnetic field at point x_j = 45 m and y_j = 1 m of Figure 20, variation of real values of X component versus real values of Y component results in a straight line for the point of consideration, x_j = 9 and $y_j = 1$ m, as shown in Figure 21, which forms the minor axis of the unmitigated magnetic field ellipse with mitigating magnetic field as its major axis, as was shown previously in Figure 11.

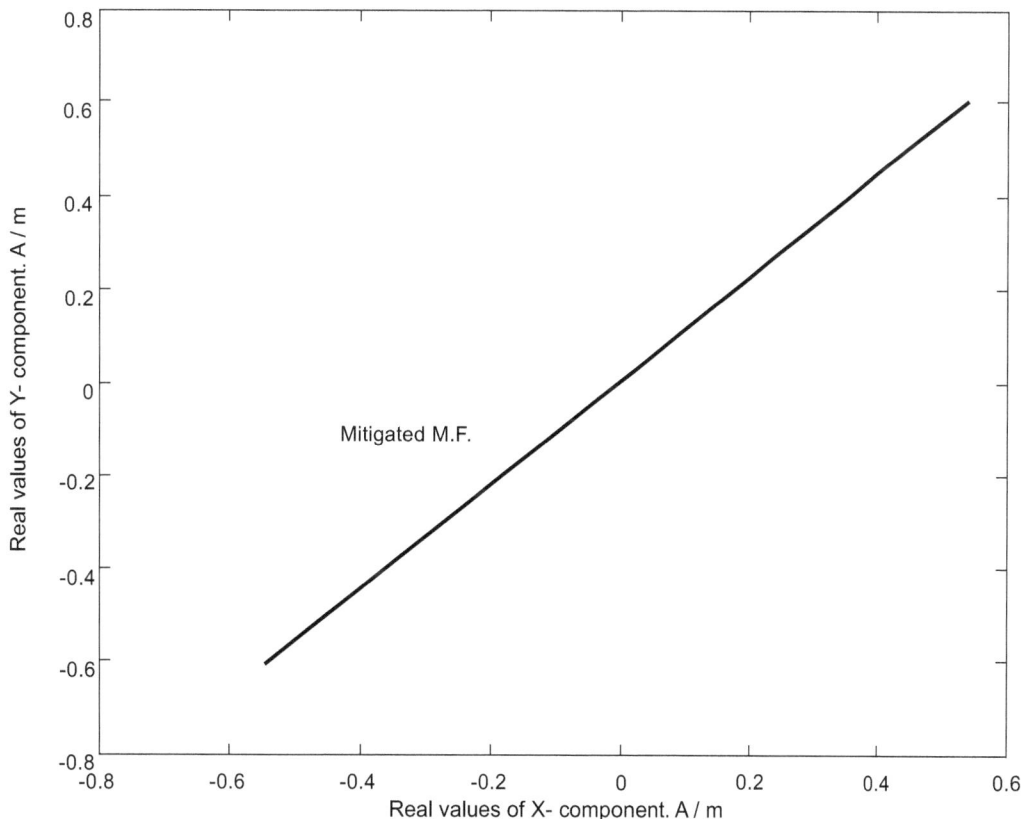

Figure 21. Position of mitigated M.F. at the point of consideration, x=9m, y=1m.

Figure 22 illustrates, unmitigated magnetic field at location $x_j = 9$ m and $y_j = 1$ m. with three different positions for altitude h, above the power lines. The resultant unmitigated magnetic field is calculated using the Equations previously thoroughly explained in the previous chapter. Implementation of real components of Equation (14) would constitute obtaining the resultant mitigating magnetic field. Subsequently, implementation of Equation (15) results in achieving the mitigated magnetic field. Irrelevant to the position of the auxiliary mitigating loop, the mitigated magnetic field always achieves a value of zero, which is a unique demonstration of the passive loop established during this course of investigation. As it was expected, a noticeable percentage of mitigations have been obtained at other locations. Effect of geometrical location of the loop on mitigated magnetic field at the other points could be observed from this Figure.

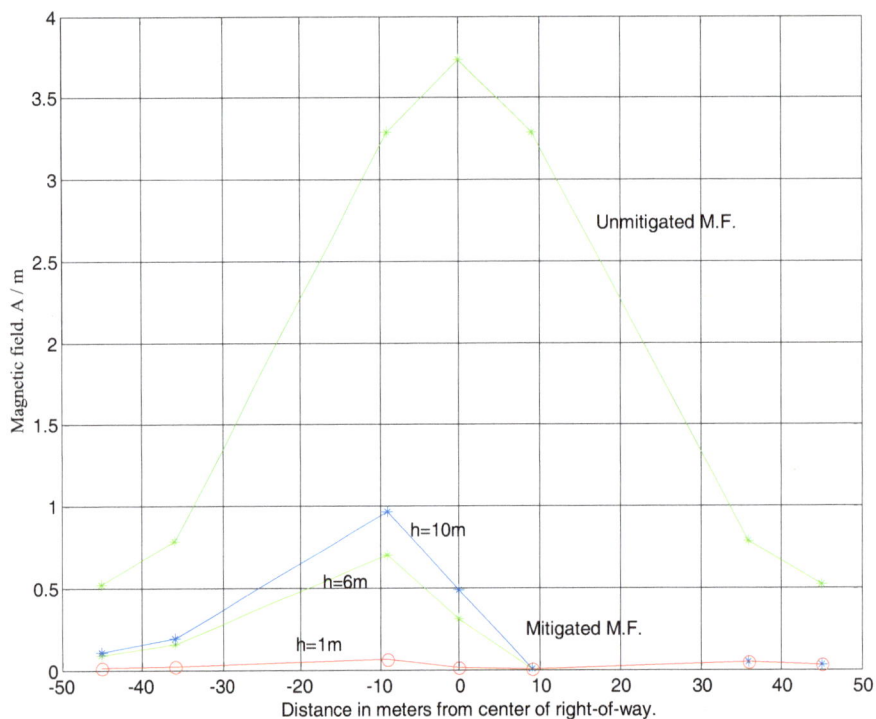

Figure 22. Relationship between unmitigated and mitigated M.F., as h varies.

As the altitude of the loop above the transmission lines increases, there would be a reduction in the induced flux in the loop, which in turn results in achieving greater amount of mitigating current in the loop as shown in Table 5.

TABLE 5. Values of currents, voltages and impedances of the loop as h varies above the power lines.

Position of h w. r. t. the power lines	Magnitude of the loop impedance. Ohms	Magnitude of the loop current. Amps.	Loop voltage. Volts
h=10 m	0.0491	883.6892	43.3896
h=6 m	0.1038	666.5044	69.1616
h=5 m	0.1282	617.5217	79.1822
h=4 m	0.1609	570.6264	91.8023
h=3 m	0.2063	525.7908	108.4594
h=2 m	0.2741	482.9833	132.3625
h =1 m	0.3929	442.1676	173.7260

At the point of consideration, $x_j = 9$ m and $y_j = 1$ m, of Figure 24 value of unmitigated magnetic field, 2.4406 - 2.2023i, is equal to that of mitigating field, -2.4406 + 2.2023i, but their orientations are in the opposite directions, resulting in a zero value of mitigated magnetic field irrelevant to the altitude of the auxiliary mitigating loop from the transmission lines. But as this Figure shows, at $x_j = -9$ m and $y_j = 1$ m, value of the mitigated magnetic field increases as this altitude increases. For altitude, h, equal to 1 m, the mitigating magnetic field would be equal to 2.4732 + 2.1655i, resulting in a noticeable reduction of the unmitigated field. Magnitude of the mitigated magnetic field for the given case would be equal to 0.0636 A / m.

At an altitude of 10 meters above the transmission line, the loop produces a mitigating magnetic field of value 1.3142 + 3.0130i. The vector sum of unmitigated and mitigating fields at this point constitutes a mitigated magnetic field of magnitude 0.9694 A /m. As Figure 24 shows, value of mitigated magnetic field for h = 6 meters above the power line falls somewhere between these two values.

Let us consider location $x_j = -9$ m, $y_j = 1$ m. Let r_1 be the distance from center of mitigating conductor M_1, which is installed above phase A of the transmission line, as shown in Figure 23.

$$r_1 = \sqrt{(h+15)^2}$$

Obviously, r_1 increases, as h is increased.

Let r_2 be distance from center of mitigating conductor M_2 to the same point. Therefore;

$$r_2 = \sqrt{(18)^2 + (15+h)^2}$$

Similarly, as h increases, r_2 increases.

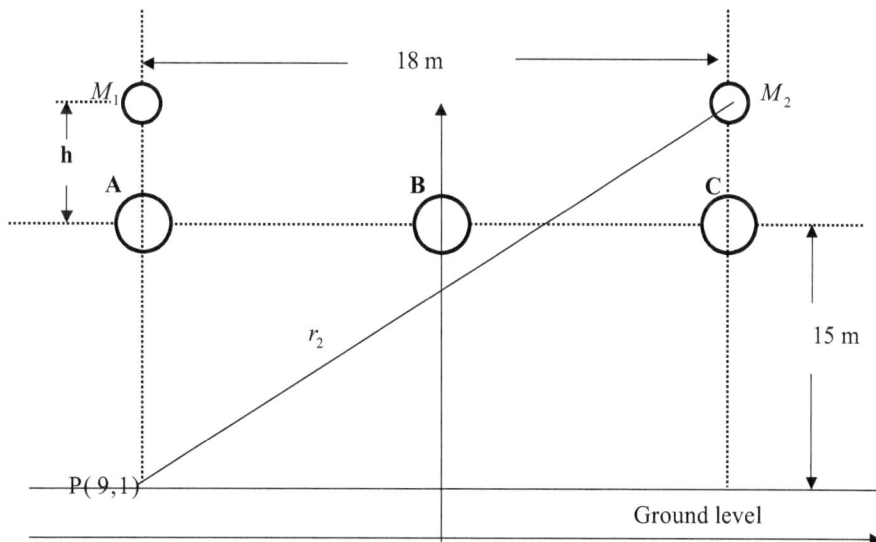

Figure 23. Schematic diagram of a three-phase line. Mitigating conductors installed above.

$$\sin\left(\beta_1\right)=0$$
$$\cos\left(\beta_1\right)=-1$$
$$\sin\left(\beta_2\right)=\frac{-18}{r_2}$$
$$\cos\left(\beta_2\right)=-\frac{15+h}{r_2}$$

Equation (16) shows that in addition to r_1 and r_2, other parameter such as $\cos\left(\beta_1\right)$, $\cos\left(\beta_2\right)$, $\sin\left(\beta_1\right)$ and $\sin\left(\beta_2\right)$ are also effective in calculating mitigating magnetic field. These parameters are, obviously, function of altitude, h. From the same Figure, sine term has only one variable.

Therefore, as h increases denominator r_2, which is function of h increases, resulting in reduction of the $\sin\left(\beta_2\right)$

Since in case of cosine term, denominator and numerator are both function of h, $\cos\left(\beta_2\right)$ increases as h increases. A close look at the Equation (16) reveals the fact that as r_1 and r_2 increase, the mitigating field, $\vec{H}_m$, decreases resulting in less magnetic field to be mitigated.

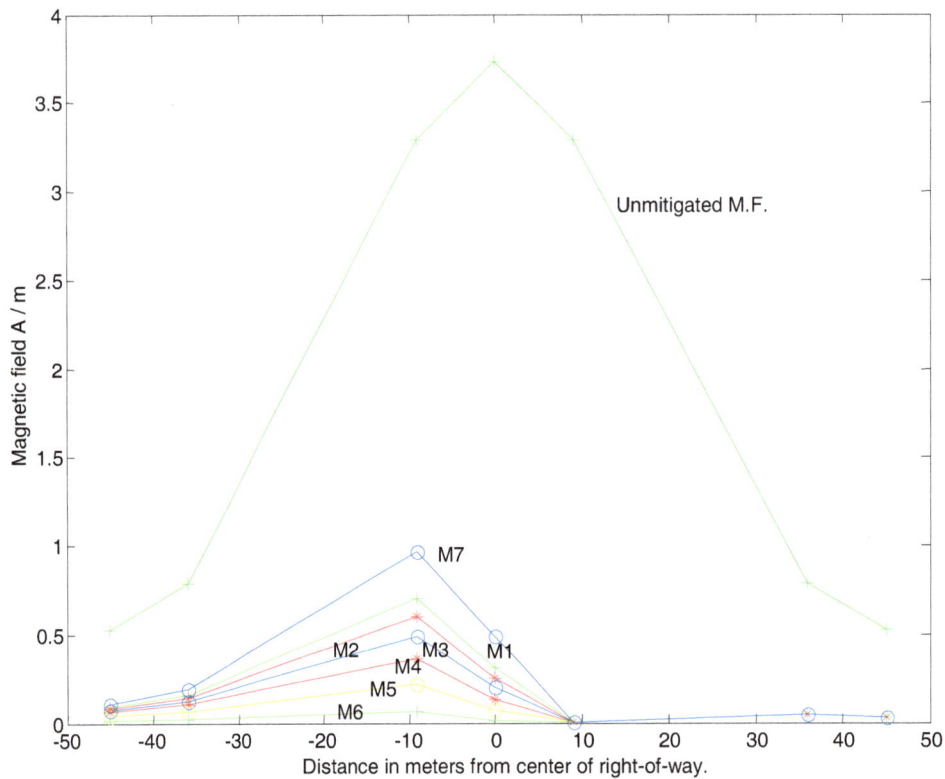

Figure 24. Variations of mitigated M.F. as h varies. Point of consideration: x=-9m, y=1m.

Figure 24 shows values of mitigated magnetic field in A / m for seven different locations with $x = 9m$ and $y = 1m$ as point of consideration, when the auxiliary mitigating loop is placed above the transmission line. As this Figure shows, reduction in unmitigated magnetic fields declines as altitude is increased from one meter to six meters, with one meter interval as indicated by M_1 to M_6 and then to an altitude of ten meters indicated by M_7.

5.5. Comparing the Two Locations

Figure 25 shows a comparison between mitigated magnetic field when the auxiliary loop is positioned above the high voltage transmission line and then beneath the line for seven different locations, namely –45, -36, -9, 0, 9, 36, and 45 meters away from the center of right-of-way. At each position, above and below, the auxiliary loop is allowed to take six different altitudes, namely from one meter to six meters, with one meter interval, from the transmission line. Letters A in this Figure indicate above and letters b indicate below.

As this Figure shows, the value of mitigated magnetic fields for locations, 45 m and 36 m from center of the right-of-way remains almost the same irrelevant to position of the auxiliary mitigating loop from power line. It can be observed that altitude of the mitigating loop cannot influence the mitigated magnetic field at these two locations. Obviously, since location $x_j = 9$ m , $y_j = 1$ m is the point of consideration, at both the positions(above and below the power lines) and at any altitude from the transmission line, hundred percent mitigation has been achieved. This Figure shows that magnitude of mitigated magnetic field is influenced by the position as well as by the altitude of the auxiliary mitigating loop from the power line and it reaches its peak value at $x_j = $ -9 m, $y_j = 1$ m, which once again justifies our previous explanation.

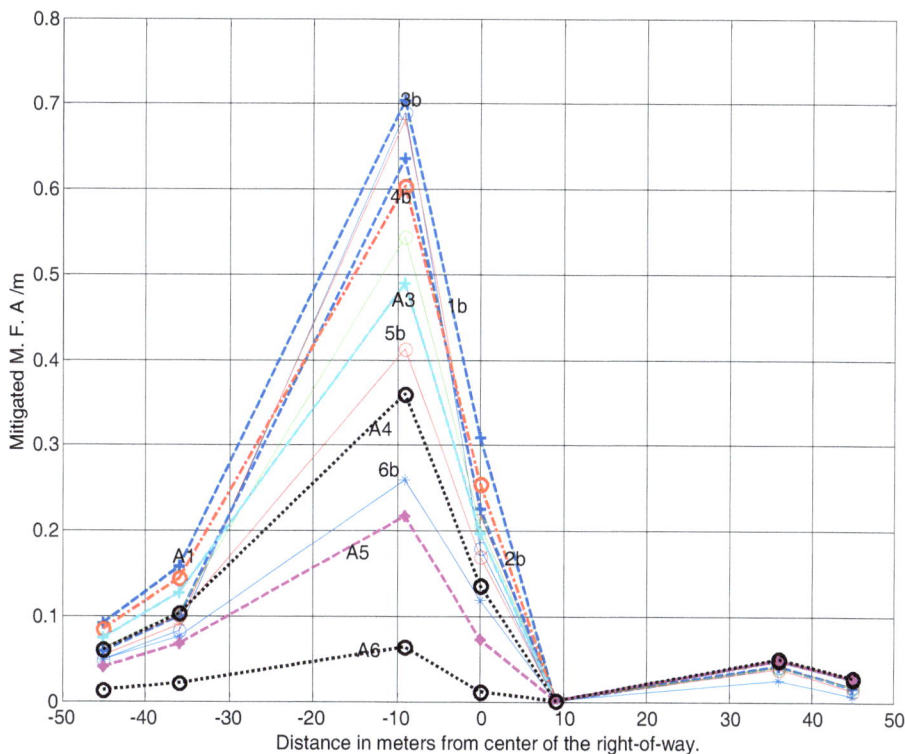

Figure 25. Characteristics of mitigated M.F. for auxiliary loop above and below the power lines.

At $x_j = 45\,\text{m}$, $y_j = 1$ m when the auxiliary mitigating loop is placed above the transmission line r_1 obtains values between 57.94 m and 56.32 m. Simultaneously, r_2 takes values between 41.69 m and 39.4 m. The mitigating current values vary from 666.5 Amps to 442.16 Amps, as the loop is allowed to select six different positions, as shown in Table 6

As the same Table illustrates, when the loop is placed below the transmission line, r_1 achieves values between 54.75 m and 55.79 m and r_2 obtains values between 37.10 m and 38.63 m. The mitigating current achieves values between 207.84 Amps and 366.34 Amps.

TABLE **6**. Values of parameters for two positions of the mitigating loop

Loop positioned below the transmission line			Loop above the line		Distance, from the line
r_1, meters	r_2, meters	I_m, Amps	r_1, meters	r_2, meters	
54.75	37.10	207.84	57.94	41.68	h = 6m
54.92	37.37	236.29	57.59	41.18	h = 5 m
55.10	37.64	266.27	57.25	40.70	h =4 m
55.32	37.95	297.89	56.92	40.25	h =3 m
55.54	38.28	331.22	56.61	39.81	h =2 m
55.79	38.63	366.34	56.32	39.39	h = 1 m

Substitution of the corresponding values of r_1 and r_2 and sine and cosine terms, which in turn are function of these two parameters, in real components of Equation (14) results in obtaining the mitigating magnetic fields of almost the same values. Such very close values of mitigating fields cause the obtained mitigated magnetic fields to be of almost the same values.

At point $x_j = -9m$ and $y_j = 1m$, influence of position (above and below) of the auxiliary mitigating loop and its locations are more visible. As Figure 25 shows, the mitigating loop produces the highest value of mitigating magnetic field, resulting in obtaining the best mitigation at this point when the loop is positioned above and at an altitude of one meter from the power line. This Figure also shows that when the auxiliary loop is positioned below and at a distance of six meters from the transmission line produces the highest value of mitigated magnetic field.

At this point the mitigated magnetic field is fully influenced by the position as well as the altitude of the loop. As this Figure shows, the loop produces the best mitigation of 0.0636 A / m, when it is located above and at a distance of one meter from the power lines. The corresponding value of the mitigated magnetic field when the mitigating loop is located below the transmission line is equal to 0.2594 A / m, roughly four times higher. At the altitude of five and six meters from the power lines, values of the mitigated magnetic fields are not much affected by the position of the loop, but as this loop gets closer to the power lines, position of the auxiliary mitigating loop above the transmission line influences value of the mitigated magnetic field.

For further understanding of these phenomena, effects of mitigating magnetic field on unmitigated field have been scrutinized.

5.6. Process of Mitigation

In order to illuminate process of mitigation, a comparative process has been established. During this process, first the mitigating loop is installed below the two outer phases of the transmission line and then above the power line.

Case 1.

5.7 Auxiliary Loop is Placed Below the Line

Figure 26 shows the characteristics of three different magnetic fields, unmitigated, mitigating and mitigated for location $x_j = -9m$ and $y_j = 1m$. The auxiliary mitigating loop is installed below the two outer phases of the transmission line.

In general, in order to achieve reduction of unmitigated magnetic field at any location, unmitigated magnetic field and mitigating magnetic field are simultaneously allowed to vary over one complete cycle of 360°. During this variation, unmitigated magnetic field vector corresponding to a specific angular frequency is vectorially added up to the mitigating magnetic field of the same angular frequency. As Figure 26 shows, at angular frequency equal to zero degree, unmitigated magnetic field vector is equal to -2.09 - 2.28i and the value of the mitigating magnetic field corresponding to the same angular frequency is equal to 2.12 + 1.41i, resulting in a mitigated magnetic field vector of + 0.03 - 0.87i, producing a mitigated magnetic field of magnitude of 0.8705 A / m. Unmitigated magnetic field reaches its maximum value of -2.4406 - 2.2023i at angular frequency of 21°. Corresponding value of mitigating magnetic field at angular frequency of 21° is equal to +2.60 + 1.73i, resulting in achieving a mitigated magnetic field vector of +0.16 – 0.47i, with a magnitude of 0.4965 A /m. Mitigating magnetic field reaches its maximum value of +2.7353 + 1.8234i at an angular frequency of almost 40°. At this angular frequency, the unmitigated magnetic field vector would be equal to –2.48 – 1.88i

Consequently, the mitigated magnetic field vector would be equal to +0.26 –0.05i with a magnitude of 0.2648 A / m. From the same Figure, unmitigated magnetic field obtains its minimum value of -0.5367 + 0.6162i, generating a magnitude of 0.8172 A / m at an angular frequency of 111°. The corresponding value of mitigating magnetic field at this angular frequency would be equal to 0.86 + 0.57i, generating a mitigated magnetic field vector of +0.32 + 1.19i, with magnitude of 1.2323 A / m.

Case 2.

5.8. Auxiliary Loop is Placed Above the Power Line

Figure 27 shows the characteristics of the three magnetic fields for location $x_j = -9m$ and $y_j = 1m$, when the loop is placed above the two outer phases of the same transmission line.

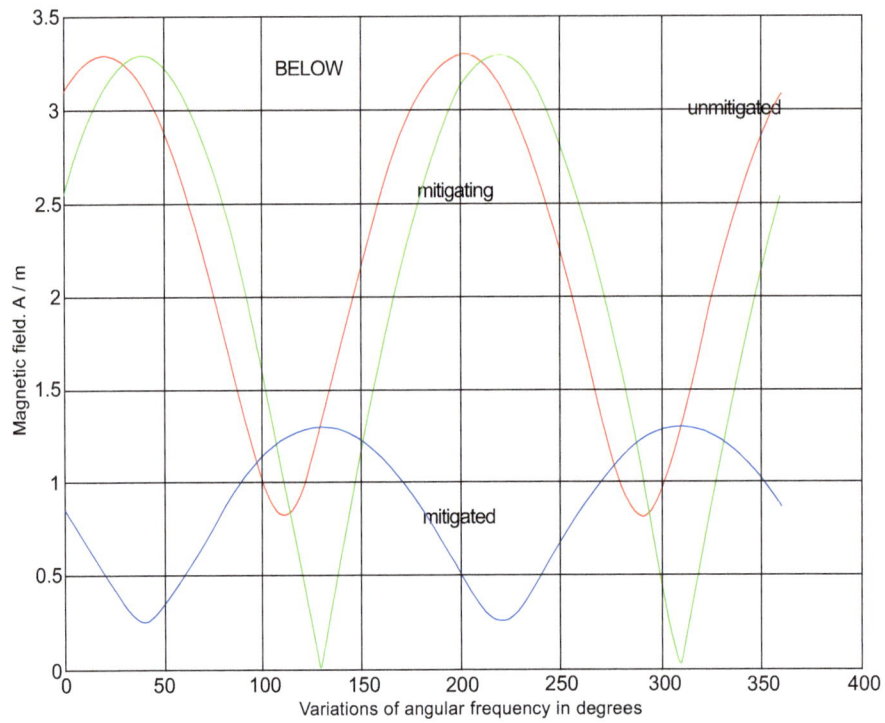

Figure 26. Three magnetic fields, the loop is below the lines, h=1m, location, x=-9 m, y=1m.

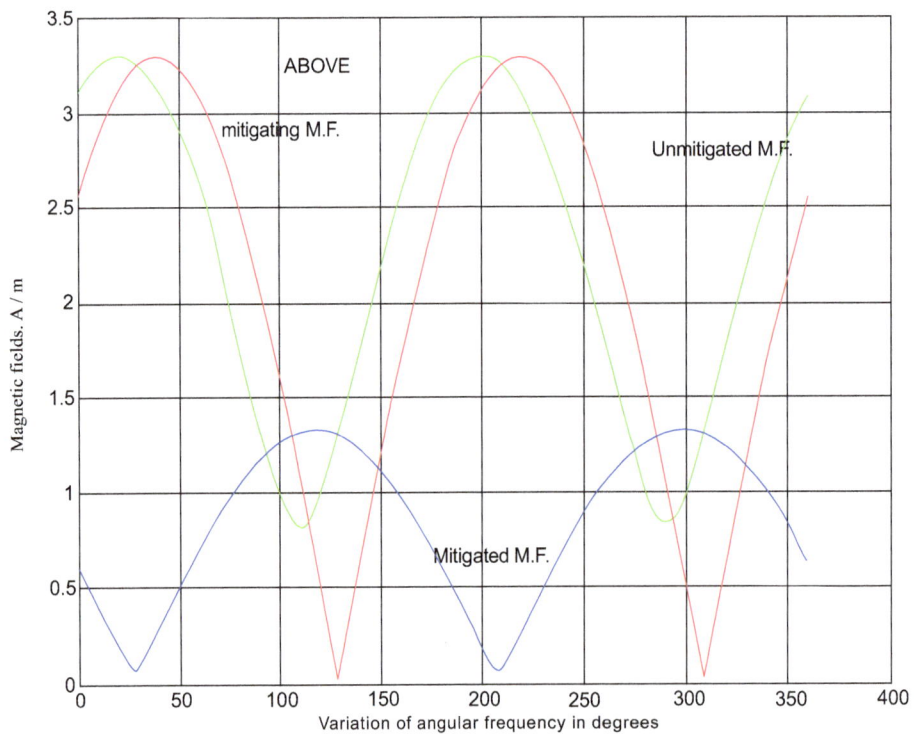

Figure 27. Position of three magnetic fields, the loop located above the lines, x=-9m, y=1m.

For further understanding the reason why when the auxiliary mitigating loop is located above the transmission line produces better mitigation than when it is installed below the line and in particular why it generates the best mitigation of magnetic field when the mitigating loop is installed at a distance of one meter above the transmission lines, as shown in Figure 25, let us scrutinize Figure 27.

As depicted in Figure 27, mitigating magnetic field obtains its maximum value of 2.4732 + 2.1655i at an angular frequency of 42° with the corresponding value of unmitigated magnetic field equals to –2.47 – 1.84i. The unmitigated magnetic field achieves its maximum value of -2.4406 - 2.2023i when angular frequency is equal to 21°. At this angular frequency, the mitigating magnetic field possesses a value of 2.35 + 2.06i, constituting a mitigated magnetic field vector of – 0.09 – 0.14i, with a magnitude of 0.1664 A / m.

The minimum mitigated magnetic field occurs at an angular frequency of 28°. At this angular frequency, unmitigated magnetic field obtains value of –2.49 – 2.11i and mitigating magnetic field is equal to 2.43 + 2.12i, resulting in a mitigated magnetic field vector of - 0.06 + 0.01i, establishing a mitigated field of magnitude 0.06 A / m, as can be observed from Figure 27.

As Figure 27 shows, mitigating magnetic field achieves it minimum value of -0.04 – 0.02i at an angular frequency of 130°. The corresponding value of unmitigated magnetic field vector at this angular frequency would be equal to + 0.29 +1.30i, constituting a mitigated magnetic field vector of 0.25 + 1.28i producing a magnitude of 1.3042 A / m.
During the second half a cycle, unmitigated magnetic field obtains its maximum value of 2.4406 + 2.2023i at 201° and its minimum value of +0.5367-0.6162i at 291°. Subsequently, mitigating magnetic field achieves its maximum value of –2.7353 – 1.8234i at an angular frequency of 220°. In order to illustrate further understanding of mitigation phenomena, let us consider the angular frequency of 95°. At this value of angular frequency, unmitigated magnetic field achieves a value of –1.19 – 0.01i and that of mitigating magnetic field would be equal to +1.54 + 1.03i. Vector sum of these two vectors results in mitigated magnetic field vector of + 0.35 + 1.02i, constituting a magnitude of 1.0784 A / m.

It is clearly to be understood that during mitigation process, values of unmitigated and mitigating magnetic fields vectors corresponding to the same angular frequency are vectorially added up to achieve the mitigated magnetic field vector during one complete cycle of angular frequency variation.

At angular frequency equal to 118°, unmitigated magnetic field vector is equal to – 0.24 + 0.88i and that of mitigating field vector is equal to 0.48 + 0.42i, resulting in a mitigated magnetic field vector of + 0.24 + 1.30i, with magnitude of 1.32 A / m, as shown in the same Figure. Please bear in mind that the entire calculations are performed using the previously explained Equations.

As these Equations show, for a fixed position of the auxiliary mitigating loop, the only parameter that remains effective is the angular frequency.

Comparing the two different positions (below and above), the minimum value of mitigated magnetic field, when the mitigating loop is installed below and at a distance of one meter occurs at an angular frequency of 40°, whereas the angular frequency at which minimum value of mitigated magnetic field when it is placed above and one meter away from the power line occurs at an angular frequency of 28°.

The second factors that should be considered are the geometrical locations, such as distances from center of the mitigating loop conductor to the location, x = -9m and y =1m. The X and Y coordinate of the loop conductor not only influence the distances r_1 and r_2, but also their effectiveness can be observed on sine and cosine terms of Equation (16).

5.9. Discussions

By now, it has been realized that it would be almost easier to install an additional loop beneath the two outer phases of a high voltage transmission line. Such an installation requires that the tower strength be carefully studied. Effect of ice as well as wind loading must also be thoroughly checked. In order to achieve a complete satisfaction of the operation, the capacitor voltage rating in the loop must be checked. The loop conductor parameters should also be carefully checked for its corona performances.

Installing the auxiliary mitigating loop above the transmission line, would be somewhat more challenging and all the conditions already mentioned are equally applicable to this case. Though, installing the auxiliary mitigating loop above the line may require to increasing the height of the tower, installation of the loop beneath the power line may cause ground clearance problem and the low voltage in the auxiliary loop particularly at the mid span must be checked. Such restrictions are not applicable for the case when the auxiliary loop is placed above the transmission line, but further percussions must be taken to protect the loop against the lightning.

In both the cases, the developed method is well applicable to mitigate the magnetic field produced by the transmission line. The established approach is well capable to create hundred percent reduction of magnetic field at any desired point and simultaneously to result in a noticeable reduction of the magnetic field at the other points. A better mitigation is achieved when auxiliary mitigating loop is positioned above the transmission line. In our case, the best mitigation is obtained when the loop is installed at one meter above the power line.

Even though, for simplicity of the calculation a mitigating loop of 1000 meters has been selected, but the developed approach can be applied to any length of the auxiliary mitigating loop. Such an alteration would not cause any changes in the mitigated field. Speed and direction of flow of wind with respect to the mitigating loop must be studied. A strong possibility exists that a disturbance in the mitigation process may be registered, when wind is gusting at a very high speed.

CHAPTER 6

Magnetic Field and Delta Connections

Abstract: In order to demonstrate the capability of the developed approach on any types of transmission line, delta configuration power line has been investigated and numerical illustration is established. The results are tabulated and the related figures are depicted. In this chapter, characteristics of the magnetic fields and effect of the altitude of the auxiliary mitigating loop with respect to the power line are studied. Finally, relationship between the three types of the magnetic fields for delta – connected configuration has been scrutinized and the depicted figure illuminates the discussions.

6.1. Delta Configuration

In order to demonstrate the capability of the established methods on other type of configuration, a delta configuration power line will be considered.

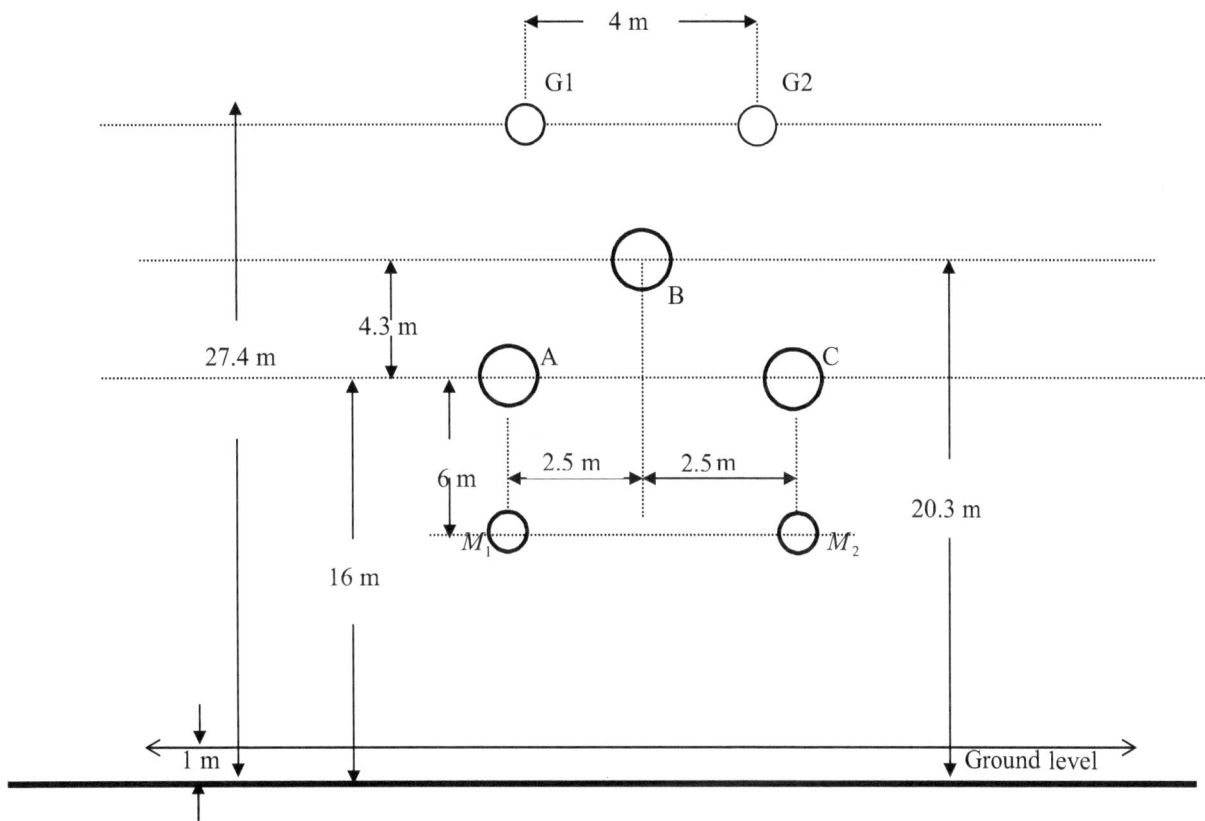

Figure 28. A delta configuration with auxiliary mitigating loop installed beneath the two phases A and C.

Figure 28 shows delta – connected configuration. Table 7 shows the angular frequencies at which maximum unmitigated magnetic fields corresponding to nine different locations occur.

TABLE **7**. Angular frequencies at which maximum unmitigated magnetic field is produced for a delta configured power line.

Distance from center of the right-of-way. Meters.	-10	-7.5	-5	-2.5	0	2.5	5	7.5	10
ωt, degrees	14.7	18	21.7	25.8	30	34.2	38.3	42	45.3

A comparison between Table 2 and Table 7 shows that at the point of consideration $x_j = 0\,\mathrm{m}$, $y_j = 1\,\mathrm{m}$, the angular frequencies, in both the cases, obtain the same value of $30°$. These angular frequencies are responsible to produce maximum unmitigated magnetic fields at those points.

In order to understand the reason, let us consider flat and delta configurations of Figures 3 and 28. It is obvious that parameters R_A, R_B and R_C, are not the same, as shown in Table 8.

TABLE **8**. Relationships between parameters for two configurations

	Flat configuration	Delta configuration
R_A	17.4929	15.2069
R_B	15	19.3000
R_C	17.4929	15.2069

Consequently, sine and cosine terms of Equation (11), in both the cases, would not be the same.

For flat configuration: For delta configuration:

$\cos(\alpha_a) = 0.8575$ $\cos(\alpha_a) = 0.9864$

$\sin(\alpha_a) = 0.5145$ $\sin(\alpha_a) = 0.1644$

$\cos(\alpha_b) = 1$ $\cos(\alpha_b) = 1$

$\sin(\alpha_b) = 0$ $\sin(\alpha_b) = 0$

$\cos(\alpha_c) = 0.8575$ $\cos(\alpha_c) = 0.9864$

$\sin(\alpha_c) = -0.5145$ $\sin(\alpha_c) = -0.1644$

Substitution of the above parameters in Equation (11) for flat configuration results in obtaining ; 0.0153 - 0.0255i as the numerator and -0.0088 - 0.0441i as denominator.

Whereas, substitution of the above parameter in Equation (11) for delta configuration results in achieving: -0.0113 - 0.0094i as numerator and 0.0065 - 0.0162i as denominator.

Even though, in these two cases values of numerators and denominators are different, but ratio of numerator to denominator in each case would be the same, constituting achieving the value of $30°$.

Figure 29 shows characteristics for the three types of magnetic fields for a delta - connected configuration as shown in Figure 28 with x=0 m, y =1 m as point of consideration.

The unmitigated magnetic field establishes an ellipse when the angular frequency is allowed to vary over one complete cycle of $360°$. This Figure shows that mitigating magnetic field forms the major axis and mitigated magnetic field shapes the minor axis of this ellipse.

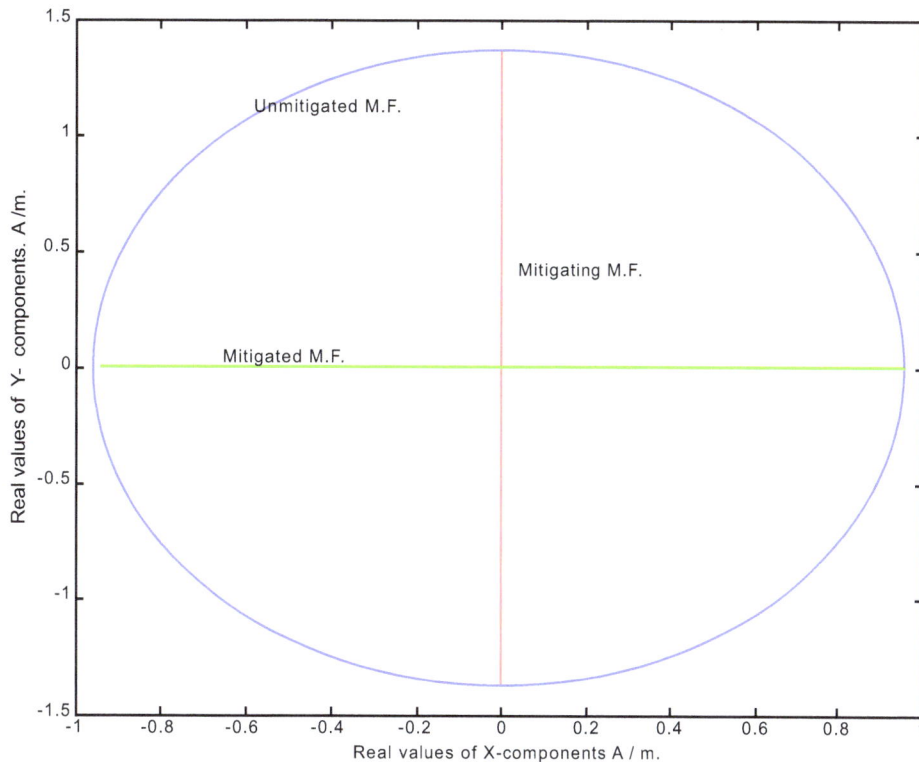

Figure 29. Characteristics of the three magnetic fields. Delta-connection, -x=0, m, y1 m.

6.2. Characteristics of the Magnetic Fields

Variation of the three types of magnetic fields over one complete cycle of 360° is depicted in Figure 30. As this Figure illustrates, the mitigated magnetic field achieves its zero values at ωt equals to 30° and ωt equals to 210° respectively. At any value of angular frequency between 30° and 210°, value of unmitigated magnetic field would not be the same as that of mitigating magnetic field. Subsequently, values of mitigated magnetic field within these values of angular frequencies would be greater than zero.

The unmitigated magnetic field fluctuates between a maximum value of 1.3709 A / m and a minimum value of 0.9555 A /m. The mitigating magnetic field also varies between a maximum value of 1.3709 A / m and a minimum value of zero as shown in Figure 30.

The mitigated magnetic field obtains its maximum value of 0.9555 A / m.

Relationship between the three magnetic fields for delta-connected power lines is exactly the same as when the three phases are having the same y coordinates (flat configuration). As Figure 30 reveals, the mitigated magnetic field after achieving its first zero value at angular frequency of 30˚, increases until it reaches its maximum value of 0.9555 A / m, which is equivalent to the minimum value of the unmitigated magnetic field, which thereafter it declines until it obtains its zero value at angular frequency of 210˚.

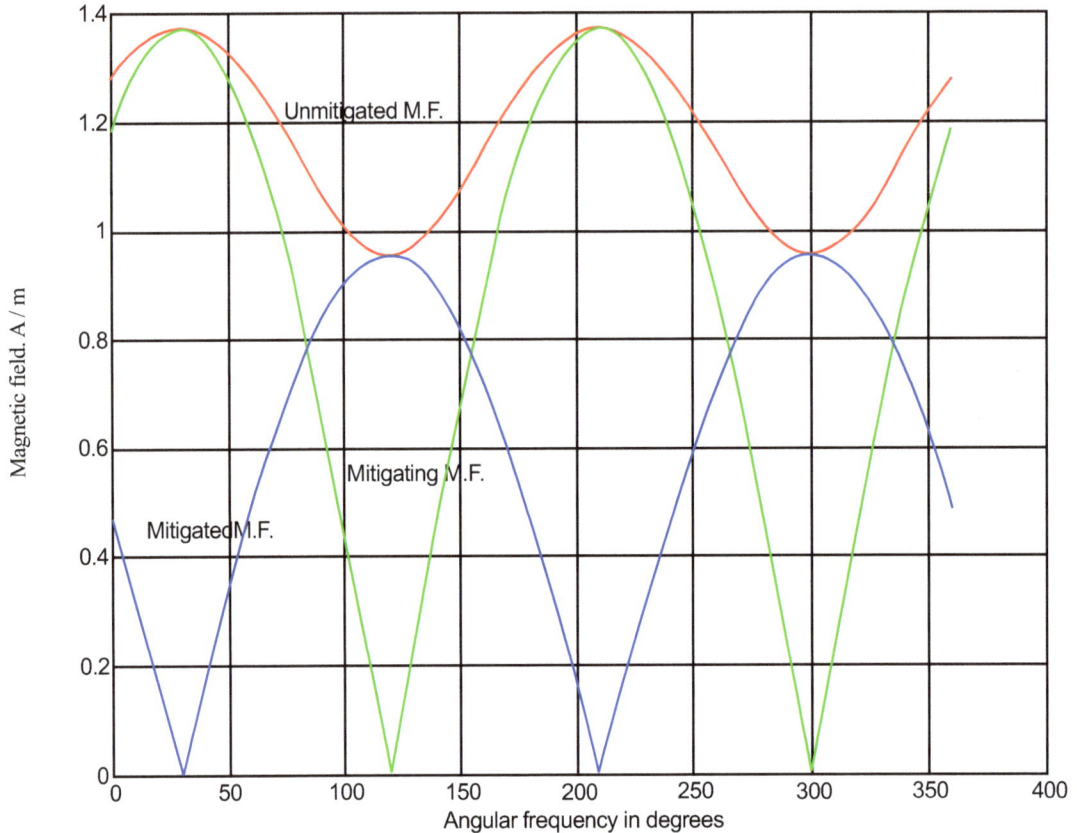

Figure 30. Relationships between the three magnetic fields .Delta connection, x=0 m, y=1 m.

Comparing the three magnetic fields of Figure 30 with one another, this Figure shows that at angular frequency of 30°, unmitigated and mitigating magnetic fields achieve their maximum values, (as it was expected). These phenomenon are the same as in the case of flat configuration, but with this exception that their numerical values differ, which is obviously due to the distances from the center of the phase conductors to the point of consideration and other factors such as the loop voltage, which in turn is a function of geometrical location of the auxiliary mitigating loop.

6.3. Effect of Altitude

Let us examine the effect of changing the altitude of the auxiliary mitigating loop from the three phases. The loop is permitted to move from a distance of six meters beneath the two outer phases of the conductors to a distance of one meter, with x_j=0 m and y_j=1 m as the point of consideration.

Let us consider the case, when the auxiliary loop is installed at a distance of six meters below the two outer phases

As Figure 31 shows, hundred percent of mitigation has been achieved at the point of consideration. Magnificent reductions of unmitigated fields have also been obtained at the other eight points.

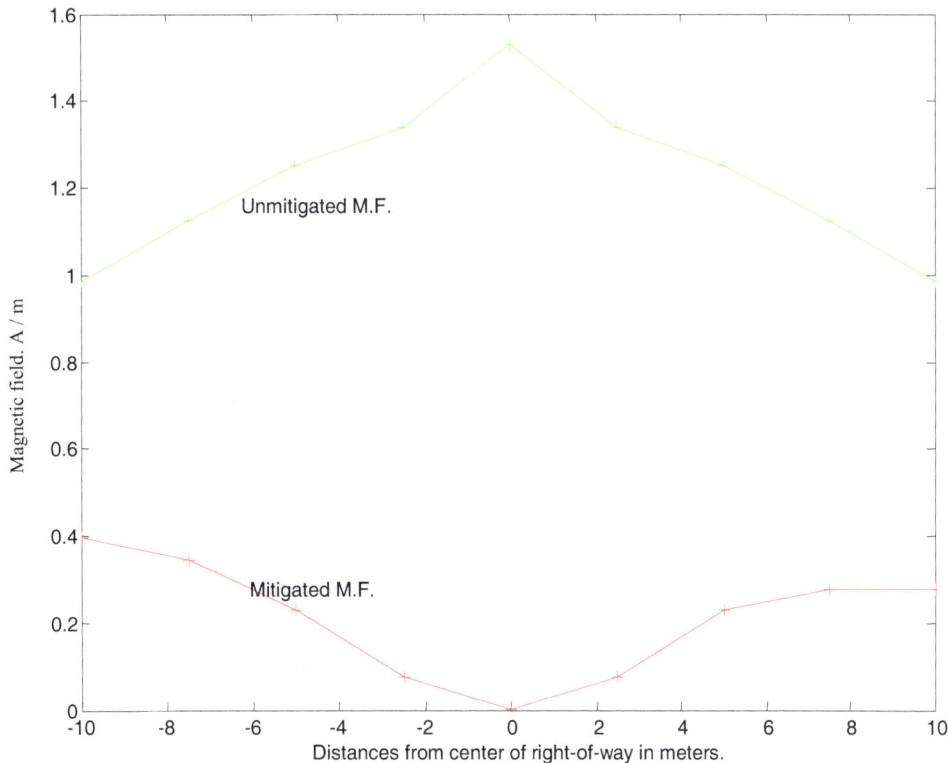

Figure 31. Relationship between unmitigated and mitigated M.F. at 9 points.

In order to further illuminate the applicability of the developed method, the auxiliary mitigating loop is then placed at a distance of one meter from the outer phases of the same conductors and the results are shown in Table 9.

TABLE **9**. The loop is installed at two positions beneath the lines. Point of consideration: x_j =0m, y_j =1m.

Distance from center of the right-of-way. Meters		-10	-7.5	-5	-2.5	0	2.5	5	7.5	10
Unmitigated M.F. A /m		0.9840	1.1239	1.2495	1.3385	1.5311	1.3385	1.2495	1.1239	0.9840
Mitigated M.F. A / m.	h=6 meters	0.3962	0.3457	0.2295	0.0782	0.000	0.0782	0.2295	0.2780	0.2771
	h=1 meter	0.0483	0.0373	0.0212	0.0079	0.000	0.0079	0.0212	0.0077	0.0082

As this Table shows, values of the mitigated magnetic fields have noticeably declined at each location, when the auxiliary mitigating loop is placed at a distance of one meter from the conductors. Therefore, a better mitigation is achievable at h = 1 m.

In addition to $\sin(\beta)$ and $\cos(\beta)$ of Equation (16) that influence the mitigating magnetic field, mitigating current is also a very influential factor. Obviously, due to geometrical effect values of cosine terms slightly reduce as the auxiliary mitigating loop is moved from h = 1 meter to h = 6 meters, but at the same time, values of the sine terms increase. Simultaneously, the value of mitigating current drops to more than half. Consequently, the amount of mitigating magnetic field produced at h = 1 meter would be much higher than when the loop is located at h = 6 meters, resulting in better mitigation at the other points as shown in Table 9

Let us consider location $x_j = 10m$ and $y_j = 1m$ (ninth entry of Table 9)

At h=6 meters   At h = 1 meter

$$\cos(\beta_1) = -0.5843$$
$$\sin(\beta_1) = -0.8115$$
$$\cos(\beta_2) = -0.7682$$
$$\sin(\beta_2) = -0.6402$$

$$\cos(\beta_1) = -0.7459$$
$$\sin(\beta_1) = -0.6660$$
$$\cos(\beta_2) = -0.8815$$
$$\sin(\beta_2) = -0.4722$$

$$I_m = -150.3070 \text{ Amps}$$ $$I_m = -348.4209 \text{ Amps}$$

mitigating magnetic field = -0.7743 + 0.0176i   mitigating field = -0.8327 + 0.4841i

6.4 Relationships Between the Magnetic Fields

In order to demonstrate capability of our findings, let us consider $x_j = 2.5$ m, $y_j = 1$ m
Figure 32, shows the characteristics of the three types of the magnetic fields, unmitigated, mitigating and mitigated magnetic fields. When the unmitigated field is allowed to vary over one complete cycle, it produces an ellipse, whose major axis is over lapped by the mitigating magnetic field, resulting in a zero mitigated magnetic field. Mitigated magnetic field shapes the minor axis of this ellipse. Consequently, Theses two axes are at right angle with respect to each other.

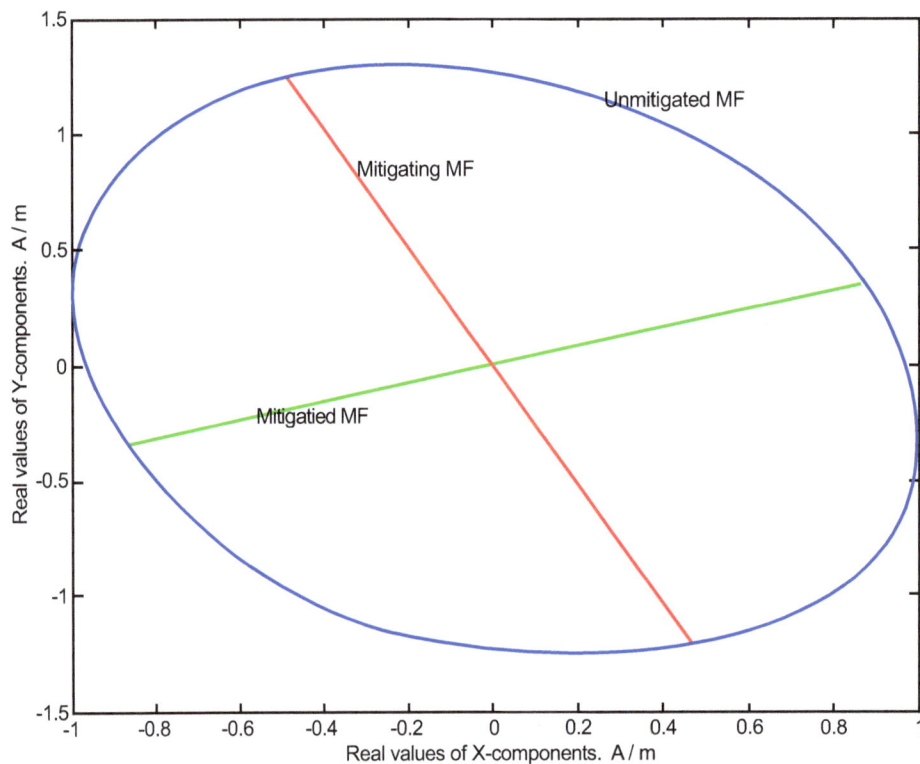

Figure 32. Characteristics of three types of MF. Point of consideration: x=2.5 m, y = 1 m.

As this Figure shows, maximum value of the unmitigated ellipse is equal to 1.34 A / m. Value of the mitigated magnetic field drops to zero at an angular frequency of 34.2 degrees, with its maximum value of 0.933 A / m, which is the same as that of minimum value of the unmitigated magnetic field, as also shown in Figure 33.

The geometrical effect on unmitigated, mitigating and mitigated magnetic fields can easily be viewed, when Figures 30 and 33 are compared.

Referring to the Figure 33, it can be observed that the maximum value of mitigated magnetic field is always equal to the minimum value of the unmitigated magnetic field. This phenomenon is true for any point of consideration. In other words, it can be said that the minimum value of the unmitigated magnetic field is always responsible to shape the maximum value of the mitigated magnetic field, for any point of consideration. Further investigations also reveal that mitigating magnetic field drops to zero at the same angular frequency at which minimum value of unmitigated magnetic field equals the maximum value of mitigated magnetic fields. At this angular frequency the mitigating magnetic field does not exert any effect on the unmitigated field allowing the mitigated magnetic field to reach its maximum value, which is equal to the minimum value of the unmitigated magnetic field.

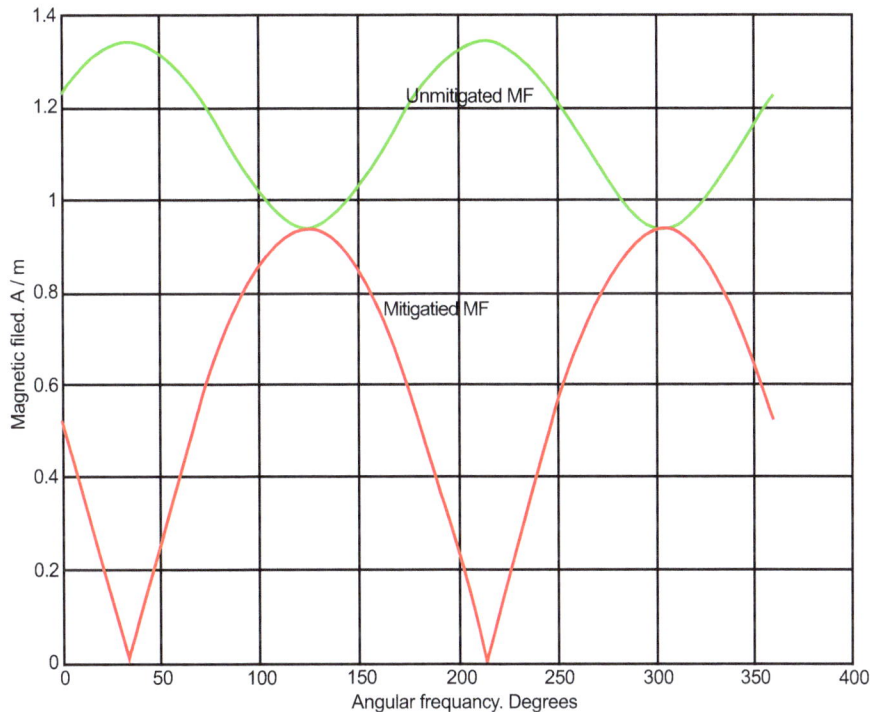

Figure 33. Relationship between unmitigated and mitigated magnetic fields. x=2.5 m, y =1 m.

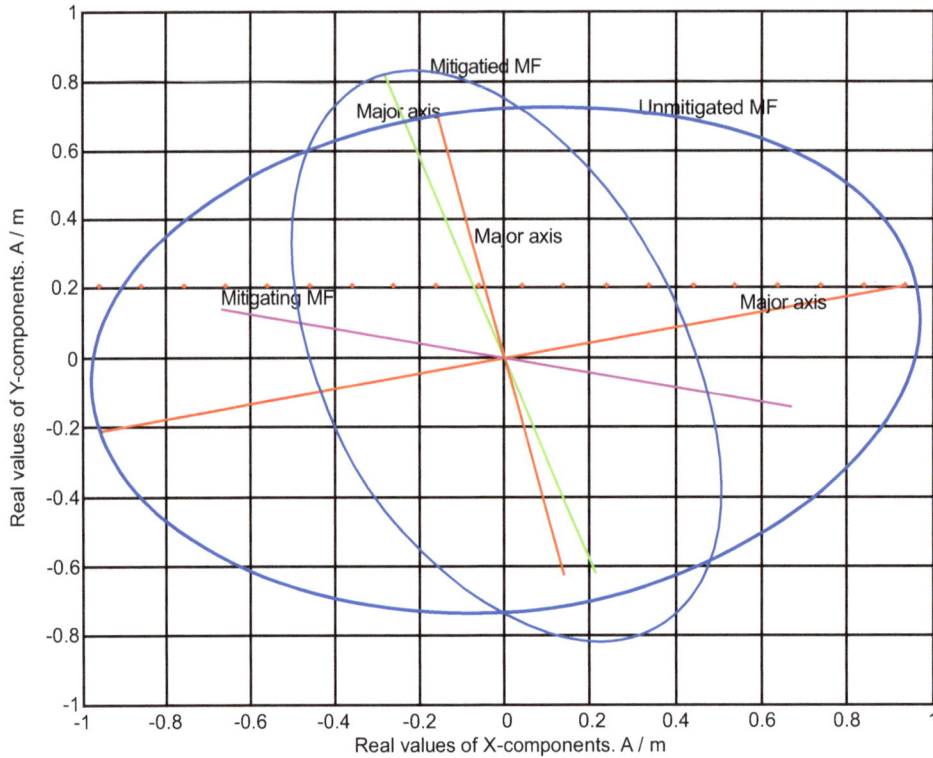

Figure 34. Position of axes and their corresponding magnetic fields.

Figure 34 shows characteristics of the three types of magnetic fields for point $x = -10m$ and $y = 1m$, with point $x = 2.5m$ and $y = 1m$ as the point of consideration for a delta - connected line as shown in Figure 28. Since it is desired to obtain hundred percent mitigation at the point of consideration, noticeable amount of mitigations are achievable at the other points.

As has already been explained, in order to achieve zero mitigated magnetic field, major axis of the unmitigated magnetic field must be equal in magnitude to that of mitigating magnetic field, with their orientations in exact opposition. These two phasors shape the major axis of the unmitigated magnetic field.

In this process, the mitigated magnetic field, which would be equal to zero only at the particular angular frequency at which the maximum unmitigated magnetic field was achievable, shapes the minor axis of the unmitigated magnetic field. This is consistence with the fact that the two axes of any ellipse are always in right angle with respect to each other.

6.5. Contribution of Mitigation on Other Points

It has been made clear that at any point of consideration, the unmitigated magnetic field is mitigated to zero. Such process also influences the magnetic fields of other locations, but they never drop to zero. For clarity of this discussion, let us investigate location $x_j = -10m$, $y_j = 1m$, where magnetic field drops from 0.9840 A / m to 0.3962 A / m, as shown in Table 9. As Figure 34 shows, the mitigating magnetic field is not overlapping the major axis of unmitigated field. Instead, it makes an angle of 24° with the major axis of the unmitigated ellipse. Consequently, the mitigated magnetic field has a value more than zero.

The trajectory of this mitigated field over one complete cycle of 360° is also depicted in the same Figure. As this Figure indicates, major axis of mitigated field ellipse produces an angle with minor axis of unmitigated field. Had the mitigated field been equal to zero, then these two axes would have overlapped

CHAPTER 7

Bundled-Conductors Magnetic Field Calculations

Abstract: In this chapter, for further illumination of the developed approach to mitigate magnetic field associated with high voltage transmission line, bundled-conductors configuration has been scrutinized.
Each sub-conductor is separately analyzed and an equation to calculate the total unmitigated magnetic field is achieved from which, angular frequency responsible to generate maximum value of unmitigated magnetic field is set. An approach to calculate the mitigating loop impedance is also established.
The applicability of the developed method has been illustrated and effect of mitigation at seven different locations within the right-of-way has been thoroughly investigated.
Process of mitigation and variation of the three types of the magnetic fields with respect to each other has been studied and the related figures and Tables are depicted.

7.1 Bundled Conductors

When voltage of a transmission line exceeds 230 KV, the effectiveness of corona becomes more if only one conductor per phase is used. It is therefore preferred to utilize more than one conductor per phase, which is known as bundling of conductors. Therefore, a bundled conductor is one, which is made of two, three or even more conductors, which are generally known as sub-conductors. These sub-conductors are placed on a perimeter of a circle called bundle circle, as shown in Figure 35.

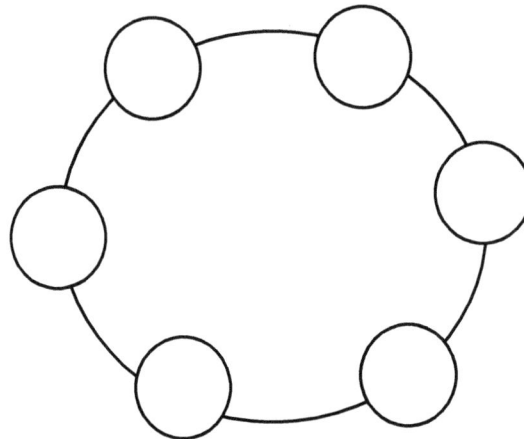

Figure 35. Bundled conductors.

The sub-conductors are placed much closer to one another as compared with the separation of the three phases.

The relationship between sub-conductor spacing S and radius of bundle circle R is given by $S = 2R\sin\left(\dfrac{\pi}{n}\right)$, where n is number of sub-conductors.

In order to achieve minimum voltage gradient on the surface of a sub-conductor, the optimum spacing between the sub-conductors must be carefully calculated, which is usually eight to ten times the diameter of the conductor. Reduction of voltage gradient results in radio interference reduction.

By bundling, GMD is obviously increased resulting in reduction of inductance L but capacitance C increases, as a result the surge impedance, which is given by $\sqrt{\dfrac{L}{C}}$ is reduced. Therefore, maximum power that can be transmitted is increased.

Not only it is economically justified to use bundled conductors, but it also reduces voltage gradient and interference with communication lines. The surge impedance is also reduced by bundling the conductors.

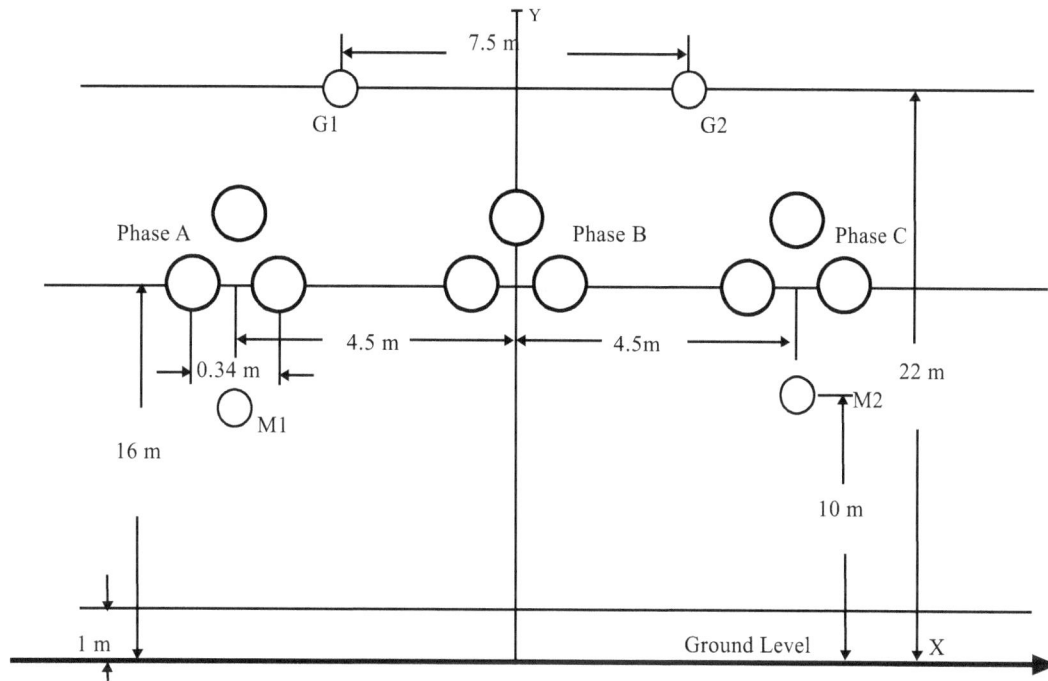

Figure 36. Bundled conductors with three sub-conductors on each phase.

In order to illuminate the effectiveness of the developed methods on the bundled conductors transmission line, a configuration as shown in Figure 36 is investigated.

As this Figure shows, each phase comprises three sub-conductors. These sub-conductors are placed at a distance of 0.34 meters from each other. The total current of 510 Amps in each phase is, obviously, equally divided among the three sub-conductors. Consequently, each of the sub-conductors carries a current of 170 Amps. It is assumed that these three phases are at 0, -120 and +120 degrees with respect to each other.

As has already been explained, the first step to calculate the maximum unmitigated magnetic field contributed by this configuration is to determine the angular frequency.

Let us consider seven different locations, such as –13.5 m, -9 m, -4.5 m, 0 m, 4.5 m, 9 m, and 13.5 m from the center of the right – of – way. It is also assumed that the object is one meter above the ground level.

7.2. Calculation of Angular Frequency

Considering each phase separately, unmitigated magnetic field produced by each sub-conductor is calculated. The vector sum of these three magnetic fields results in obtaining the magnetic field produced by each phase, at any point in the space. Then, the magnetic field produced by each phase is vectorially added up, which results in obtaining the total unmitigated magnetic field produced by this configuration.

Let the unmitigated magnetic field produced by the first sub-conductor of phase A be $\bar{H}_{a1}$, then

$$\bar{H}_{a1} = \left[\frac{I_{a1} \cos(\omega t)}{2\pi r_{a1}} \right] (\cos(\alpha_{a1}) + i * \sin(\alpha_{a1}))$$

Similarly, the unmitigated magnetic field $\bar{H}_{a2}$ and $\bar{H}_{a3}$ produced by the second and third sub-conductor are given by;

$$\bar{H}_{a2} = \left[\frac{I_{a2} \cos(\omega t)}{2\pi r_{a2}} \right] (\cos(\alpha_{a2}) + i * \sin(\alpha_{a2})) (27)$$

$$\bar{H}_{a3} = \left[\frac{I_{a3} \cos(\omega t)}{2\pi r_{a3}} \right] (\cos(\alpha_{a3}) + i * \sin(\alpha_{a3}))$$

The three sub-conductors of phase B contribute unmitigated magnetic field as given by (28)

$$\bar{H}_{b1} = \left[\frac{I_{b1} \cos(\omega t - 120°)}{2\pi r_{b1}} \right] (\cos(\alpha_{b1}) + i * \sin(\alpha_{b1}))$$

$$\bar{H}_{b2} = \left[\frac{I_{b2} \cos(\omega t - 120°)}{2\pi r_{b2}} \right] (\cos(\alpha_{b2}) + i * \sin(\alpha_{b2})) (28)$$

$$\bar{H}_{b3} = \left[\frac{I_{b3} \cos(\omega t - 120°)}{2\pi r_{b3}} \right] (\cos(\alpha_{b3}) + i * \sin(\alpha_{b3}))$$

Contribution of the three sub-conductors of phase C are given by (29)

$$\bar{H}_{c1} = \left[\frac{I_{c1} \cos(\omega t + 120°)}{2\pi r_{c1}} \right] (\cos(\alpha_{c1}) + i * \sin(\alpha_{c1}))$$

$$\bar{H}_{c2} = \left[\frac{I_{c2} \cos(\omega t + 120°)}{2\pi r_{c2}} \right] (\cos(\alpha_{c2}) + i * \sin(\alpha_{c2})) (29)$$

$$\bar{H}_{c3} = \left[\frac{I_{c3} \cos(\omega t + 120°)}{2\pi r_{c3}} \right] (\cos(\alpha_{c3}) + i * \sin(\alpha_{c3}))$$

The total unmitigated magnetic field produced by phase A is given by:

$$\bar{H}_A = \bar{H}_{a1} + \bar{H}_{a2} + \bar{H}_{a3} \quad (30)$$

Similarly;

$$\bar{H}_B = \bar{H}_{b1} + \bar{H}_{b2} + \bar{H}_{b3}$$
$$\bar{H}_C = \bar{H}_{c1} + \bar{H}_{c2} + \bar{H}_{c3} \quad (31)$$

Where $\bar{H}_B$ and $\bar{H}_C$ are the unmitigated magnetic fields produced by phases B and C respectively.

The total unmitigated magnetic field produced at the point of consideration is given by

$$\bar{H}_T = \bar{H}_A + \bar{H}_B + \bar{H}_C \quad (32)$$

Substitute for the corresponding values of $\bar{H}_A$, $\bar{H}_B$ and $\bar{H}_C$ from Equations (30) and (31). Take derivative of the obtained Equation with respect to ωt and then equate it to zero.

After some manipulations:

$$\omega t = a\tan\left(-\frac{M}{K}\right)(33)$$

where;

$$
\begin{aligned}
K = {} & \frac{I_{b1}}{4*\pi}\left[\frac{\cos(\phi_{b1})}{r_{b1}} + \frac{\sin(\phi_{b1})}{r_{b1}} + \frac{\cos(\phi_{b2})}{r_{b2}} + \frac{\sin(\phi_{b2})}{r_{b2}} + \frac{\cos(\phi_{b3})}{r_{b3}} + \frac{\sin(\phi_{b3})}{r_{b3}}\right] \\
& + \frac{I_{c1}}{4*\pi}\left[\frac{\cos(\phi_{c1})}{r_{c1}} + \frac{\sin(\phi_{c1})}{r_{c1}} + \frac{\cos(\phi_{c2})}{r_{c2}} + \frac{\sin(\phi_{c2})}{r_{c2}} + \frac{\cos(\phi_{c3})}{r_{c3}} + \frac{\sin(\phi_{c3})}{r_{c3}}\right](34) \\
& - \frac{I_{a1}}{2*\pi}\left[\frac{\cos(\phi_{a1})}{r_{a1}} + \frac{\sin(\phi_{a1})}{r_{a1}} + \frac{\cos(\phi_{a2})}{r_{a2}} + \frac{\sin(\phi_{a2})}{r_{a2}} + \frac{\cos(\phi_{a3})}{r_{a3}} + \frac{\sin(\phi_{a3})}{r_{a3}}\right]
\end{aligned}
$$

and

$$
\begin{aligned}
M = {} & \frac{0.866*I_{b1}}{2*\pi}\left[\frac{\cos(\phi_{b1})}{r_{b1}} + \frac{\sin(\phi_{b1})}{r_{b1}} + \frac{\cos(\phi_{b2})}{r_{b2}} + \frac{\sin(\phi_{b2})}{r_{b2}} + \frac{\cos(\phi_{b3})}{r_{b3}} + \frac{\sin(\phi_{b3})}{r_{b3}}\right] \\
& - \frac{0.866*I_{c1}}{2*\pi}\left[\frac{\cos(\phi_{c1})}{r_{c1}} + \frac{\sin(\phi_{c1})}{r_{c1}} + \frac{\cos(\phi_{c2})}{r_{c2}} + \frac{\sin(\phi_{c2})}{r_{c2}} + \frac{\cos(\phi_{c3})}{r_{c3}} + \frac{\sin(\phi_{c3})}{r_{c3}}\right](35)
\end{aligned}
$$

7.3. Numerical Illustrations

In order to establish numerical illustration and also applicability of the developed Equations, let us consider Figure 36 and select seven different locations, -13.5 m, -9 m, -4.5 m, 0 m, 4.5 m, 9 m, 13.5 m, with object one meter above the ground level.

7.4. Angular Frequency

For the illustrative purpose of calculating angular frequency, only one location, $x_j = 0\,\text{m}$ and $y_j = 1\,\text{m}$ has been scrutinized, as depicted in Figure 36-a

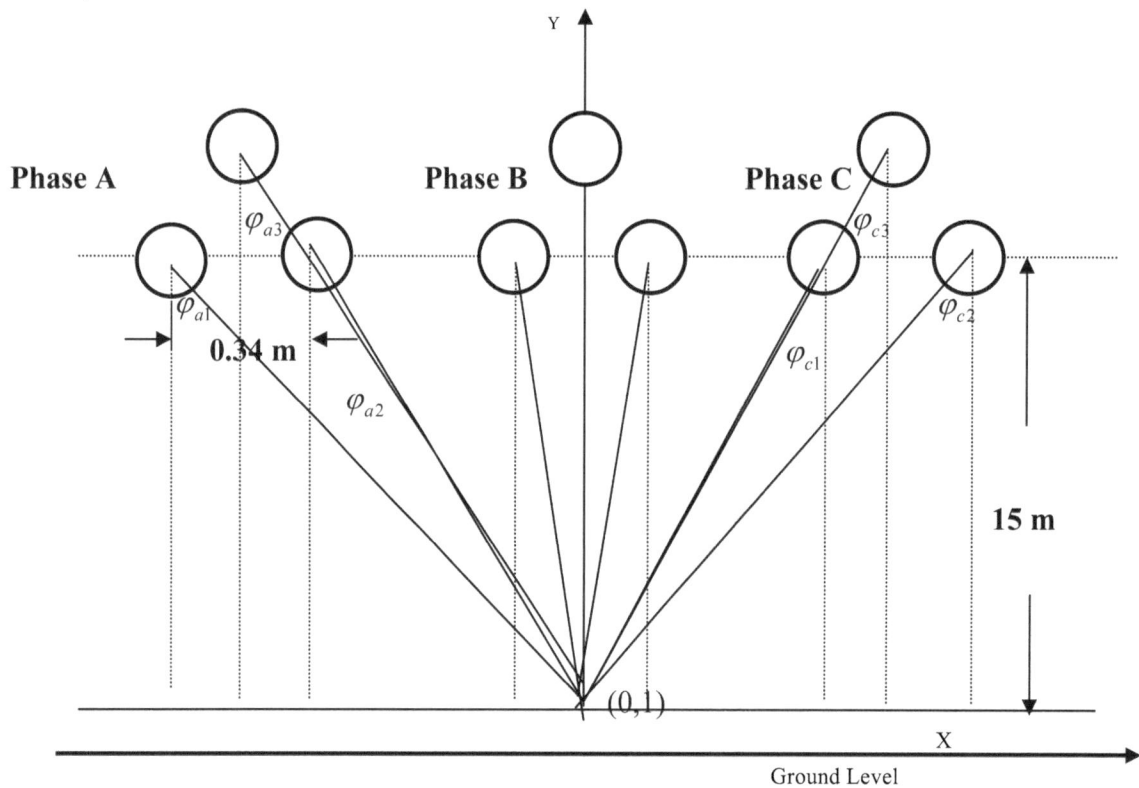

Figure 36-a. Position of the three sub-conductors with respect to the point of consideration.

From Figure 36-a,

$$r_{a1} = sqrt\left((15)^2 + (4.67)^2\right)$$
$$r_{a2} = sqrt\left((15)^2 + (4.33)^2\right)$$
$$r_{a3} = sqrt\left((15.2944)^2 + (4.5)^2\right)$$

$$\cos\left(\phi_{a1}\right) = \frac{-15}{r_{a1}}$$

$$\sin\left(\phi_{a1}\right) = \frac{-4.67}{r_{a1}}$$

$$\cos\left(\phi_{a2}\right) = \frac{-15}{r_{a2}}$$

$$\sin\left(\phi_{a2}\right) = \frac{-4.33}{r_{a2}}$$

$$\cos\left(\phi_{a3}\right) = \frac{-15.2944}{r_{a3}}$$

$$\sin\left(\phi_{a3}\right) = \frac{-4.5}{r_{a3}}$$

Considering phase B and then phase C, from the geometry of the same Figure;

$$r_{b1} = sqrt\left(\left(0.17\right)^2 + \left(15\right)^2\right)$$

$$r_{b2} = sqrt\left(\left(15\right)^2 + \left(0.17\right)^2\right)$$

$$r_{b3} = sqrt\left(\left(0\right)^2 + \left(15.2944\right)^2\right)$$

$$\cos\left(\phi_{b1}\right) = \frac{-15}{r_{b1}}$$

$$\sin\left(\phi_{b1}\right) = \frac{-0.17}{r_{b1}}$$

$$\cos\left(\phi_{b2}\right) = \frac{-15}{r_{b2}}$$

$$\sin\left(\phi_{b2}\right) = \frac{0.17}{r_{b2}}$$

$$\cos\left(\phi_{b3}\right) = -1$$

$$\sin\left(\phi_{b3}\right) = 0$$

$$\cos\left(\phi_{c1}\right) = \frac{-15}{r_{c1}}$$

$$\sin\left(\phi_{c1}\right) = \frac{4.33}{r_{c1}}$$

$$\cos\left(\phi_{c2}\right) = \frac{-15}{r_{c2}}$$

$$\sin\left(\phi_{c2}\right) = \frac{4.67}{r_{c2}}$$

$$\cos\left(\phi_{c3}\right) = \frac{-15.2944}{r_{c3}}$$

$$\sin\left(\phi_{c3}\right) = \frac{4.5}{r_{c3}}$$

Where ;

$$r_{c1} = sqrt\left(\left(15\right)^2 + \left(4.33\right)^2\right)$$

$$r_{c2} = sqrt\left(\left(15\right)^2 + \left(4.67\right)^2\right)$$

$$r_{c3} = sqrt\left(\left(15.2944\right)^2 + \left(4.5\right)^2\right)$$

Implementing MatLab programming,

M = -0.3798 - 1.2744i
K = -0.2193 + 2.2074i

Substituting for M and K in Equation (33), ωt would be equal to 30 degrees, as shown in Table 10.

Table **10**. Relationship between distance and angular frequency

Distance from center of right-of-way, meters	Angular frequencies, degrees
-13.5	25.0553
-9	25.5587
-4.5	27.1833
0	29.9992
4.5	32.8151
9	34.4398
13.5	34.9433

7.5 Calculation of Loop Voltage

As Figure 36 shows, the mitigating auxiliary loop is placed at six meters below the third sub-conductors of phases A and C. The length of this mitigating loop is assumed to be equal to 1000 meters. The developed method is well capable to produce hundred percent cancellation with any length of the auxiliary mitigating loop. The width of this mitigating loop is, obviously, 9 meters, as can be observed from Figure 36.

In order to calculate the total flux penetrating through this mitigating loop, the flux produced by each sub-conductor of each phase is determined and then vectorially added up.

Let us apply Equation (20) and calculate the flux penetrated through the mitigating loop by each sub-conductor of phase A. Each sub-conductor of phase A carries a current of magnitude 170 Amps.

$I_{a1} = 170$ Amps
$I_{a2} = 170$ Amps
$I_{a3} = 170$ Amps

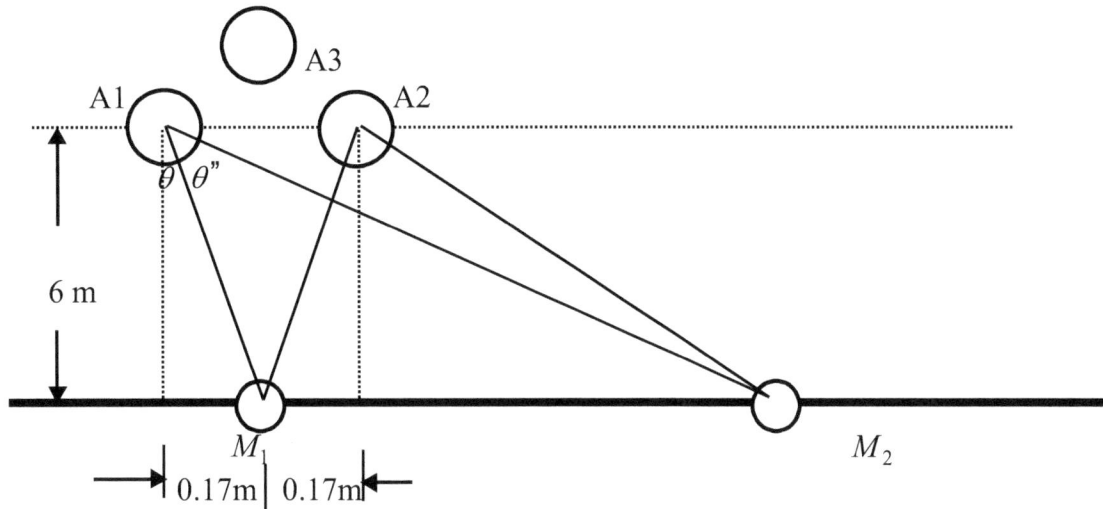

Figure 37. Position of the mitigating loop, $M_1 M_2$, with respect to sub-conductors of phase A.

From Figure 37;

First let us consider the sub-conductor A_1

$$\theta' = a\tan\left(\frac{0.17}{6}\right) = 0.0283$$

$$\theta'' = a\tan\left(\frac{9.17}{6}\right) = 0.9914$$

Substituting the above values in Equation (20);

$$\Phi_{a1} = 2*10^\wedge-7*170*\left[\ln\left(\cos\left(0.0283\right)\right) - \ln\left(\cos\left(0.9914\right)\right)\right]$$
Therefore

$$\Phi_{a1} = 2.0466e^{-005}$$

Considering the second sub-conductor A_2 of phase A;

$$\theta' = a\tan\left(\frac{0.17}{6}\right) = 0.0283$$

$$\theta'' = a\tan\left(\frac{8.83}{6}\right) = 0.9740$$

$$\Phi_{a2} = 2*10^\wedge-7*170*\left[\ln\left(\cos\left(0.0283\right)\right) - \ln\left(\cos\left(0.9740\right)\right)\right]$$

therefore;

$$\Phi_{a2} = 1.9577e^{-005}$$

Similarly, considering the third sub-conductor, A_3, of phase A;

$$\Phi_{a3} = 1.8927e^{-005}$$

The flux, Φ_a, produced by phase A would be equal to vector sum of these three fluxes.

$$\vec{\Phi}_a = \vec{\Phi}_{a1} + \vec{\Phi}_{a2} + \vec{\Phi}_{a3}$$

Therefore;

$$\vec{\Phi}_a = 5.8970e^{-005}$$

Following the similar procedures, fluxes produced by phase B and phase C are determined.

Therefore;

$$\vec{\Phi}_T = 8.8455e^{-005} - i * 5.1070e^{-005}$$

magnitude of this flux would be equal to $1.0214e^{-004}$

Consequently, the loop voltage is given by;

$$V_{loop} = 2000 * \pi * 60 * 1.0214e^{-004} = 38.5056 \text{ volts.}$$

7.6. Unmitigated Magnetic Field

In case of a bundled conductor system, each sub-conductor produces its own unmitigated magnetic field at the point of consideration, which is related to the geometrical location of the sub-conductor with respect to the point of consideration. Vector sum of unmitigated magnetic field produced by each sub-conductor of the corresponding phase results in achieving the unmitigated magnetic field produced by that phase. The total unmitigated field would be equal to the vector sum of unmitigated fields produced by the three phases.

Let $\vec{H}_{a1}$ be the unmitigated magnetic field contributed by the first sub-conductor of phase A. $\vec{H}_{a1}$ is, obviously, directly proportional to the current , but inversely proportional to its distance from the point in the space. Therefore;

$$\vec{H}_{a1} = \left[\frac{I_{a1} * \cos(\omega t)}{2 * \pi * r_{a1}} \right] \left[\cos(\alpha_{a1}) + \sin(\alpha_{a1}) \right] (36)$$

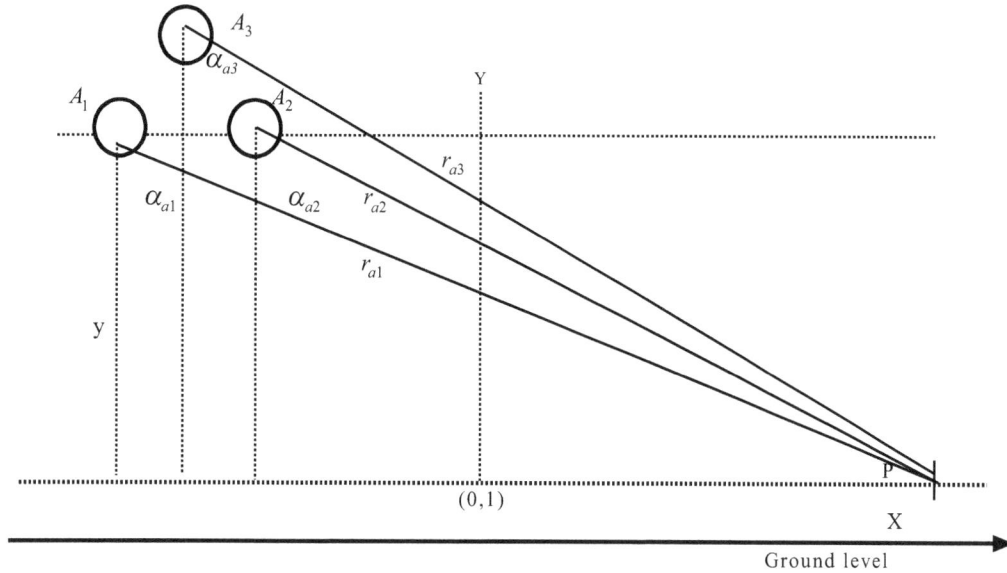

Figure 38. Geometrical position of sub-conductors of phase A with respect to a point in the space.

Where r_{a1} is distance from the center of the first sub-conductor to the point of consideration.

Similarly;

$$\vec{H}_{a2} = \left[\frac{I_{a2} * \cos(\omega t)}{2 * \pi * r_{a2}} \right] \left[\cos(\alpha_{a2}) + \sin(\alpha_{a2}) \right]$$

$$\vec{H}_{a3} = \left[\frac{I_{a3} * \cos(\omega t)}{2 * \pi * r_{a3}} \right] \left[\cos(\alpha_{a3}) + \sin(\alpha_{a3}) \right] (37)$$

α_{a_2} and α_{a_3} are the angles that r_{a_2} and r_{a_3} make with the vertical lines, as shown in Figure 38. I_{a1}, I_{a2}, I_{a3} are currents in the three sub-conductors. The total unmitigated magnetic field produced by phase A would be the vector sum of these three unmitigated fields.

$$\vec{H}_{aT} = \vec{H}_{a1} + \vec{H}_{a2} + \vec{H}_{a3} (38)$$

The three unmitigated magnetic fields produced by the three sub-conductors of phase B are given by (39).

$$\vec{H}_{b1} = \left[\frac{I_{b1} \cos(\omega t - 120°)}{2 * \pi * r_{b1}} \right] [\cos(\alpha_{b1}) + \sin(\alpha_{b1})]$$

$$\vec{H}_{b2} = \left[\frac{I_{b2} \cos(\omega t - 120°)}{2 * \pi * r_{b2}} \right] [\cos(\alpha_{b2}) + \sin(\alpha_{b2})] (39)$$

$$\vec{H}_{b3} = \left[\frac{I_{b3} \cos(\omega t - 120°)}{2 * \pi * r_{b3}} \right] [\cos(\alpha_{b3}) + \sin(\alpha_{b3})]$$

Where $\vec{H}_{bT}$, the total unmitigated magnetic field produced by phase B is given by (40).

$$\vec{H}_{bT} = \vec{H}_{b1} + \vec{H}_{b2} + \vec{H}_{b3} (40)$$

Similarly, the fields produced by the three sub-conductors of phase C are given by (41)

$$\vec{H}_{c1} = \left[\frac{I_{c1} \cos(\omega t + 120°)}{2 * \pi * r_{c1}} \right] [\cos(\alpha_{c1}) + \sin(\alpha_{c1})]$$

$$\vec{H}_{c2} = \left[\frac{I_{c2} \cos(\omega t + 120°)}{2 * \pi * r_{c2}} \right] [\cos(\alpha_{c2}) + \sin(\alpha_{c2})] (41)$$

$$\vec{H}_{c3} = \left[\frac{I_{c3} \cos(\omega t + 120°)}{2 * \pi * r_{c3}} \right] [\cos(\alpha_{c3}) + \sin(\alpha_{c3})]$$

Where ;

$$\vec{H}_{cT} = \vec{H}_{c1} + \vec{H}_{c2} + \vec{H}_{c3} (42)$$

$\vec{H}_{cT}$ is the total unmitigated magnetic field produced by phase C.

Subsequently, the total unmitigated magnetic field produced by this transmission line is the vector sum of the three fields produced by the three phases, as given by (43).

$$\vec{H}_T = \vec{H}_{aT} + \vec{H}_{bT} + \vec{H}_{cT} (43)$$

7.7. Numerical Illustrations

For the numerical illustration purpose, let us investigate location $x_j = 9$ m, $y_j = 1$ m as the point of consideration. Let us also consider ωt to being equal to 34.4398 degrees. This is the particular value of angular frequency producing maximum unmitigated magnetic field at the point of consideration.

First, let us calculate the unmitigated field contributed by each sub-conductor of phase A.

From Figure 36;

$$r_{a1} = sqrt\left(\left(13.67\right)^2 + \left(15\right)^2\right) = 20.2946$$

$$I_{a1} * \cos(\omega t) = 170 * \cos(0.6011) = 140.2014 \text{ Amps}$$

$$\cos(\alpha_{a1}) = \frac{-15}{r_{a1}}$$

$$\sin(\alpha_{a1}) = \frac{-13.67}{r_{a1}}$$

Substituting these values in Equation (36);

$$\bar{H}_{a1} = \left[\frac{140.2014}{2 * \pi * 20.2946}\right]\left(\left(\frac{-15}{r_{a1}}\right) + i * \left(\frac{-13.67}{r_{a1}}\right)\right)$$

$$\bar{H}_{a1} = -0.8127 - 0.7406i \text{ A / m}$$

Similar procedure is followed and all the parameters of Equation (37) are determined. Therefore;

$$\bar{H}_{a2} = -0.8312 - 0.7386i \quad \text{A / m}$$
$$\bar{H}_{a3} = -0.8200 - 0.7238i \quad \text{A / m}$$

Consequently, $\bar{H}_{aT}$ of Equation (38) would be equal to;

$$\bar{H}_{aT} = -2.4639 - 2.2031i \text{ A / m}$$

Now, let us consider the three sub-conductors $b_1, b_2 and b_3$ of phase B and determine all the parameters as shown in Equation (39). Implementation of this Equation results in

$$\bar{H}_{b1} = -0.1016 - 0.0621i \quad \text{A / m}$$
$$\bar{H}_{b2} = -0.1037 - 0.0610i \quad \text{A / m}$$

$$\bar{H}_{b3} = -0.1017 - 0.0599i \quad \text{A / m}$$

The total unmitigated magnetic field contributed by phase B to the point of consideration would be equal to the vector sum of these three unmitigated fields. Therefore, $\bar{H}_{bT}$ of Equation (40) would be equal to -0.3071 - 0.1830i A / m.

Similarly, $\bar{H}_{cT}$ of Equation (42) would be equal to 4.4542 + 1.3276i A / m.

The unmitigated magnetic field produced by the three phases of the considered configuration, which is depicted in Equation (43) would be equal to

$$\bar{H}_T = 1.6833 - 1.0585i \quad A/m.$$

7.8. Mitigating Loop Impedance

In order to achieve mitigation of unmitigated magnetic field contributed by the three phases of the bundled conductor transmission line, as shown in Figure 36, a mitigating loop is installed beneath the third sub-conductors of phases A and C. This loop has a width of 9 meters and is installed 6 meters beneath the line.

As the first step, from Equation (43) maximum value of the total unmitigated magnetic field at the point of consideration is calculated. For the given case, the point of consideration has coordinate of $x_j = 9m$, $y_j = 1m$ and maximum value of total unmitigated magnetic field is equal to 1.6833-i*1.0585 A / m. The corresponding value of the angular frequency, ωt, to produce this value of magnetic field is equal to 34.4398 degrees, as depicted in Table 10.

In order to achieve hundred percent mitigation, the mitigating magnetic field produced by the auxiliary mitigating loop must have a magnitude equal to that of the unmitigated magnetic field. The orientations of these two magnetic fields must be in opposition.

$$\left[\frac{I_m \cos(\omega t + \gamma)}{2*\pi*r_1}\right][\cos(\beta_1)+\sin(\beta_1)]-\left[\frac{I_m \cos(\omega t + \gamma)}{2*\pi*r_2}\right][\cos(\beta_2)+\sin(\beta_2)]=-\bar{H}_T \quad (44)$$

Where, I_m is the mitigating current.

$\gamma = -\omega t$ and

$\bar{H}_T$ is the maximum value of unmitigated magnetic field produced by the three phases of the transmission line.

Let Z_m be optimum value of the auxiliary mitigating loop impedance and V_m be the loop voltage. Therefore;

$$I_m = \frac{V_m}{Z_m} \quad (45)$$

Substituting for value of I_m from Equation (45) in Equation (44), after some manipulations;

$$Z_m = \frac{V_m \cos(\omega t + \gamma) * \left[\frac{1}{r_1}(\cos(\beta_1) + \sin(\beta_1)) - \frac{1}{r_2}(\cos(\beta_2) + \sin(\beta_2)) \right]}{-2\pi\bar{H}_T} \quad (46)$$

For the given case:

$\bar{H}_T = 1.6833 - i*1.0585$ A / m

$\omega t = 34.4398$ degrees

$V_m = 38.5056$ volts

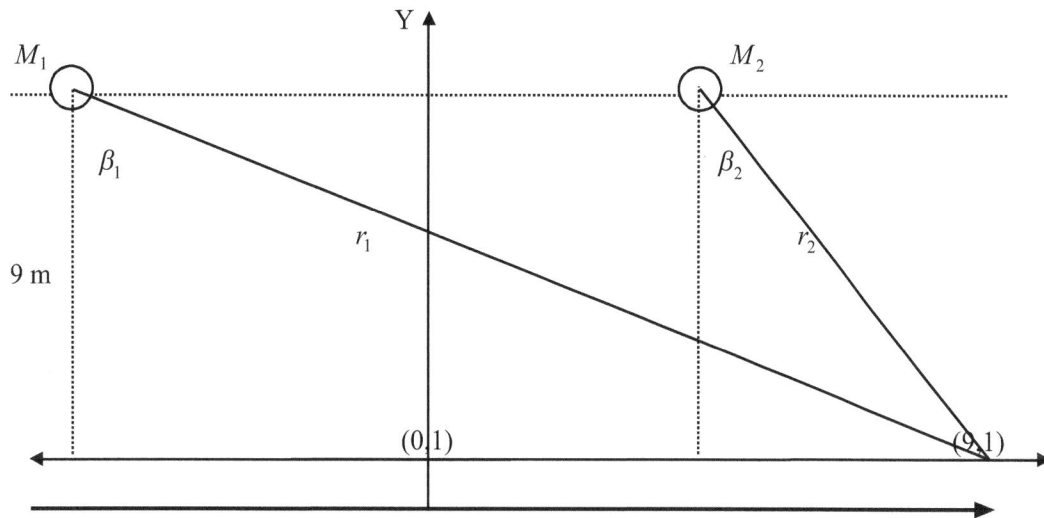

Figure 39. Position of auxiliary mitigating loop $M_1 M_2$ with respect to the point of consideration .

From Figure 39;

$$r_1 = sqrt\left((9)^2 + (13.5)^2 \right) = 16.2250$$
$$r_2 = sqrt\left((4.5)^2 + (9)^2 \right) = 10.0623$$

$$\cos(\beta_1) = \frac{-9}{r_1}$$

$$\sin(\beta_1) = \frac{-13.5}{r_1}$$

$$\cos(\beta_2) = \frac{-9}{r_2}$$

$$\sin(\beta_2) = \frac{-4.5}{r_2}$$

Substituting the above values in Equation (46),

$$Z_m = -0.1539 - 0.0719i = 0.1699 \angle 25^\circ \text{ ohms}$$

7.9. Process of Mitigation

Computation of magnetic field, which is caused by a current in a conductor, follows the Ampere's Law. This magnetic field is directly proportional to the current, but it is inversely proportional to the distance. Mitigating loop carrying a current I_{m1} could also generate magnetic field at point P located at r_1 from the conductor as depicted in Equation (47).

$$\vec{H}_{m_1} = \left[\frac{\bar{I}_{m1}}{2\pi r_1}\right]\vec{\Phi}_1 \quad (47)$$

Where ;

$$\bar{I}_{m1} = I_m \cos(\omega t + \gamma)$$

and $\vec{\Phi}_1 = [\cos(\beta_1) + \sin(\beta_1)]$

Similarly, the mitigating magnetic field produced by the second conductor is given by Equation (48)

$$\vec{H}_{m2} = \left[\frac{\bar{I}_{m2}}{2\pi r_2}\right]\vec{\Phi}_2 \quad (48)$$

Where;

$$\vec{\Phi}_2 = [\cos(\beta_2) + \sin(\beta_2)]$$

The total mitigating magnetic field contributed by these two conductors is given by (49).

$$\vec{H}_{mT} = \left[\frac{I_{m1}\cos(\omega t + \gamma)}{2\pi r_1}\right][\cos(\beta_1) + \sin(\beta_1)] - \left[\frac{I_{m2}\cos(\omega t + \gamma)}{2\pi r_2}\right][\cos(\beta_2) + \sin(\beta_2)] \quad (49)$$

In the given case;

$$\cos(\beta_1) = \frac{-9}{r_1}$$

$$\sin(\beta_1) = \frac{-13.5}{r_1}$$

$$\cos(\beta_2) = \frac{-9}{r_2}$$

$$\sin(\beta_2) = \frac{-4.5}{r_2}$$

$$\gamma = -\omega t$$

$$r_1 = sqrt\left((9)^2 + (13.5)^2\right)$$

$$r_2 = sqrt\left((4.5)^2 + (9)^2\right)$$

Substitution of the above values in Equation (49) results in achieving a mitigating magnetic field of value: -1.6833+i*1.0585 A / m.

Comparing this value of mitigating magnetic field with that of unmitigated magnetic field, one can easily observe that they have the same magnitude with their orientation in the opposite directions. The vector sum of these two magnetic fields results in zero mitigated magnetic field at the point of consideration.

7.10. Variation of the Fields

Let the currents in the sub-conductors of all the three phases vary over one complete cycle of 360° and the total unmitigated magnetic field as expressed by Equation (43) is registered. Simultaneously, variation of the mitigating magnetic field, as shown by Equation (49), over one complete cycle is also recorded.

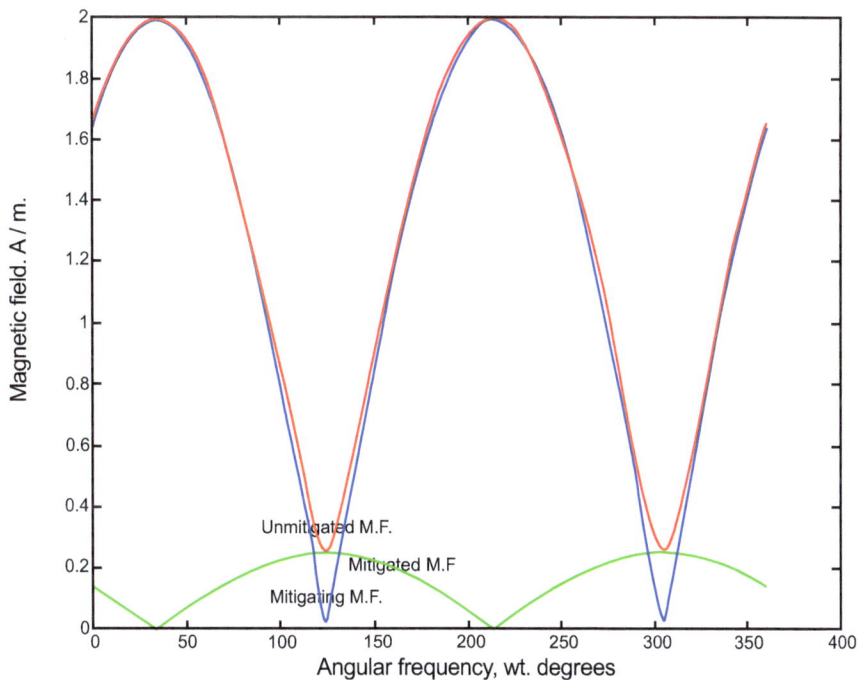

Figure. 40. Relationship between three types of magnetic fields. Point of consideration (9,1).

Figure 40 shows these variations along with the mitigated magnetic field. As this Figure shows, the unmitigated magnetic field obtains its maximum value at angular frequencies of 34.4 and 214.4 degrees respectively.

Simultaneously, at these two angular frequencies the mitigated magnetic field achieves its zero values. As depicted in the same Figure, mitigating magnetic field is almost over lapping the unmitigated magnetic field. Such phenomenon has contributed to achieving very low values of mitigated magnetic fields at other angular frequencies. The values of mitigated field vary from 0.2510 A / m to zero.

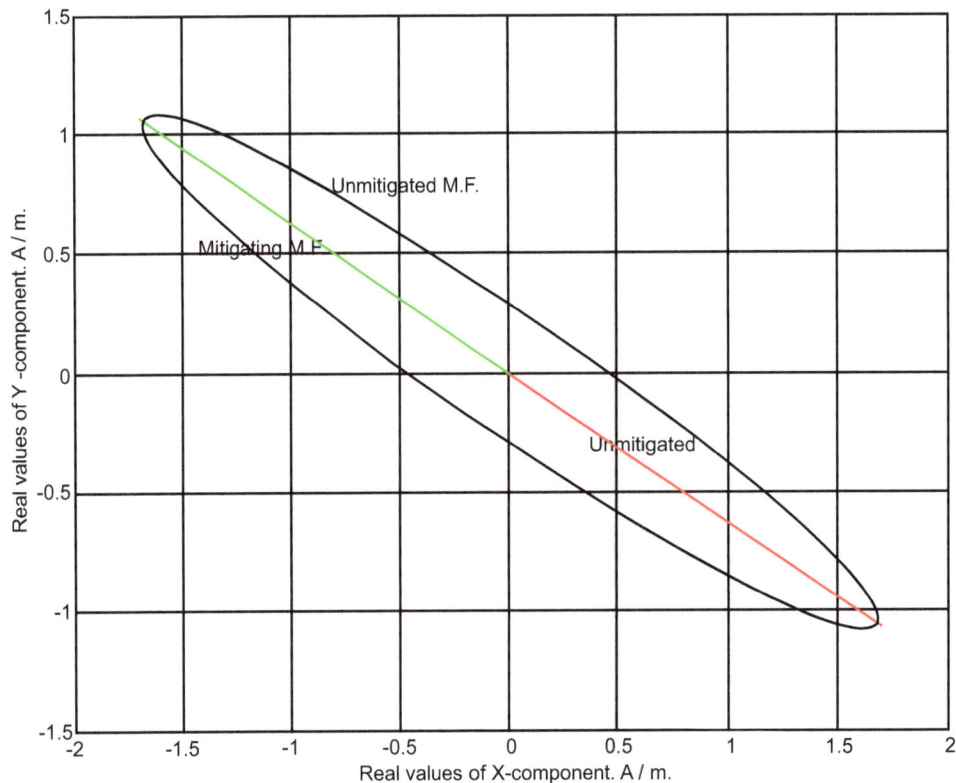

Figure 41. Characteristics of the three magnetic fields. Point of consideration (9,1).

Figure 41 shows variation of real values of x-component versus real values of Y-component, resulting in an elliptical trajectory. This is the merit of the developed approach that is irrelevant to type of configuration of the transmission line, sinusoidal variation of unmitigated magnetic field always results in an elliptical trajectory. This Figure also shows that unmitigated magnetic field possesses the same magnitude as that of mitigating magnetic field, but with their orientations in the opposite directions, resulting in a zero mitigated magnetic field.

7.11. Effect of Mitigating Field on Other Locations

In order to illuminate the effectiveness of the developed approach on other locations within the right-of-way, seven different locations, -13.5, -9, -4.5, 0, 4.5, 9, 13.5 meters, have been selected.

During this process, the unmitigated magnetic fields of the above mentioned locations are allowed to vary over one complete cycle of 360 degrees. Equation (49) is well applicable to calculate the mitigating magnetic field at each location.

The mitigating current, which influences the mitigating field at all the selected locations is one which is contributed by location $x_j = 9m$, $y_j = 1m$, the point of consideration.

The vector sum of mitigating magnetic field and unmitigated field at a specified location results in achieving the mitigated magnetic field.

7.12 Numerical Illustrations

For the purpose numerical demonstration, location $x_j = 13.5m, y_j = 1m$ is investigated. All the parameters of Equations (36), (37), (39) and (41) are determined.

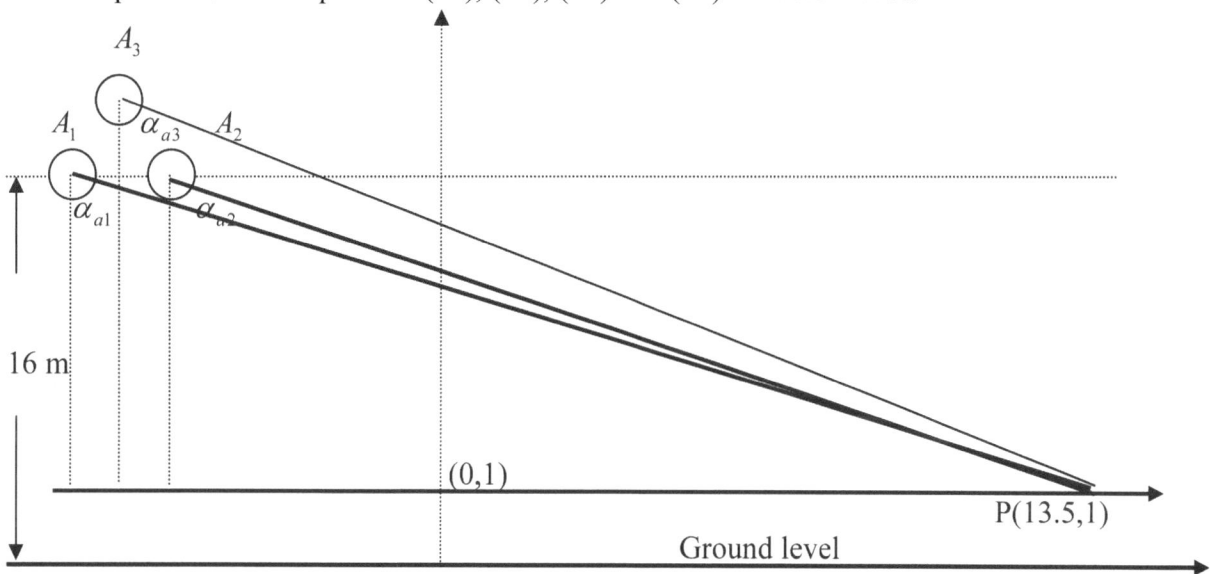

Figure 42. Position of sub-conductors of phase A with respect to the selected location (13.5,1).

From Figure 42;

$$\cos(\alpha_{a1}) = \frac{-15}{r_{a1}}$$

$$\sin(\alpha_{a1}) = \frac{-18.17}{r_{a1}}$$

$$\cos(\alpha_{a2}) = \frac{-15}{r_{a2}}$$

$$\sin(\alpha_{a2}) = \frac{-17.83}{r_{a2}}$$

Where, r_{a1}, r_{a2}, r_{a3} are equal to 23.5616, 23.3004 and 23.6203 meters respectively.

Let ωt =34.4398 degrees. This is the value of angular frequency that produces maximum unmitigated magnetic field

From Equation (36)

$$\bar{H}_{a1} = \left[\frac{170 * \cos(34.4398 * pi/180)}{2 * \pi * 23.5616}\right]\left[\frac{-15}{23.5616} + i * \frac{-18.17}{23.5616}\right] = -0.6029 - 0.7303i$$

and from Equation (37)

$$\bar{H}_{a2} = \left[\frac{170 * \cos(34.4398 * pi/180}{2 * \pi * 23.3004}\right]\left[\frac{-15}{23.3004} + i * \frac{-17.83}{23.3004}\right] = -0.6165 - 0.7328i$$

Similar procedure is followed and

$$\bar{H}_{a3} = -0.6117 - 0.7199i$$

Implementing Equation (38)

$$\bar{H}_{aT} = -1.8311 - 2.1831i$$

Following the similar approach and implementing Equations (39), (40),(41) and (42)

$$\bar{H}_{bT} = -0.2313 - 0.1548i$$

$$\bar{H}_{cT} = 3.5784 + 2.1331i$$

The total unmitigated magnetic field contributed at x_j=13.5, y_j = 1 by these three phases is given by (43). Therefore:

$$\bar{H}_T = 1.5160 - 0.2048i$$

In order to calculate the mitigating magnetic field, Equation (49) is well applicable.

$$\bar{H}_{mT} = \left[\frac{I_{m1}\cos(\omega t + \gamma)}{2\pi r_1}\right][\cos(\beta_1) + \sin(\beta_1)] - \left[\frac{I_{m2}\cos(\omega t + \gamma)}{2\pi r_2}\right][\cos(\beta_2) + \sin(\beta_2)](49)$$

From the Figure (43):

$$r_1 = sqrt\left((9)^2 + (18)^2\right)$$

$$r_2 = sqrt\left((9)^2 + (9)^2\right)$$

$$\cos(\beta_1) = \frac{-9}{r_1}$$

$$\sin(\beta_1) = \frac{-18}{r_1}$$

$$\cos(\beta_2) = \frac{-9}{r_2}$$

$$\sin(\beta_2) = \frac{-9}{r_2}$$

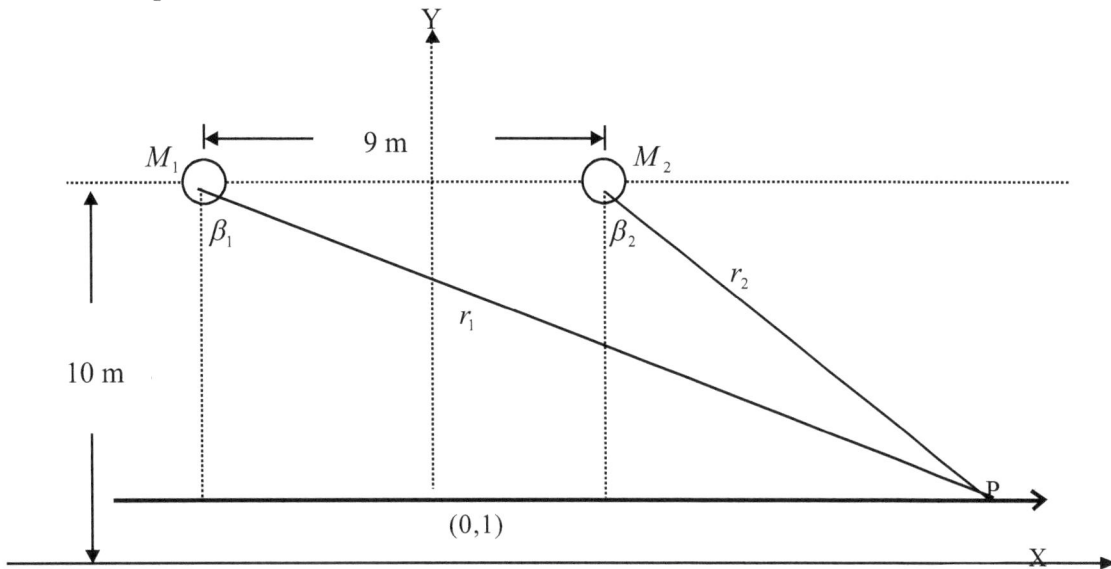

Figure 43. Geometrical position of mitigating loop with respect to point P.

Therefore;

$$\bar{H}_{mT} = -1.2591 + 0.1445i$$

The vector sum of unmitigated and mitigating magnetic fields produces the mitigated magnetic field of magnitude 0.2639 A /m.

TABLE **11**. Relationship between the two magnetic fields and distance from center of right-of-way

Distance from center of right-of- way. Meters	-13.5	-9	-4.5	0	4.5	9	13.5
Unmitigated M.F.	1.5173	1.9652	2.3683	2.5415	2.3856	1.9885	1.5298
Mitigated M.F.	1.2451	1.6363	1.7697	1.3698	0.6401	0.0000	0.2639

Table 11 shows seven values of unmitigated and mitigated magnetic fields corresponding to the seven different locations. These values are related to the angular frequency of 34.4398 degrees, for which only location (9,1) that is considered as the point of consideration can achieve its maximum value.

Let the angular frequency ωt vary over one complete cycle of 360°. The values of maximum unmitigated and mitigated magnetic fields corresponding to these seven locations are shown in Table 12. Comparing Table 11 with Table 12, it can be observed that values of unmitigated magnetic fields in the two cases belonging to the same location are almost the same, but values of mitigated magnetic fields are of much different.

Table 10 shows the angular frequencies corresponding to these seven locations. These angular frequencies are responsible to produce the maximum values of unmitigated magnetic fields at the respective locations. As this Table reveals, these values are close to one another. Therefore, when angular frequency of 34.4 degrees is assigned to produce the unmitigated magnetic fields at the seven locations, the obtained values are very close to the maximum values of unmitigated field, when the angular frequencies of Table 10 are utilized.

In order to establish a clarity of this case, location $x_j = -13.5m$, $y_j = 1m$ has been scrutinized. The value of unmitigated magnetic field corresponding to angular frequency of 34.4 degrees is equal to −1.4996 − 0.2311i. Whereas, the value of unmitigated magnetic field with angular frequency equal to 25 degrees, as depicted in Table 10, would be equal to -1.52 − 0.26i.

The mitigating magnetic field corresponding to angular frequency of 34.4 degrees would be equal to 0.9266 − 0.8711i, resulting in achieving a mitigated magnetic field of -0.5794 − 1.1022i. The mitigating magnetic field possesses a value of 0.91 − 0.86i at angular frequency of 25 degrees.

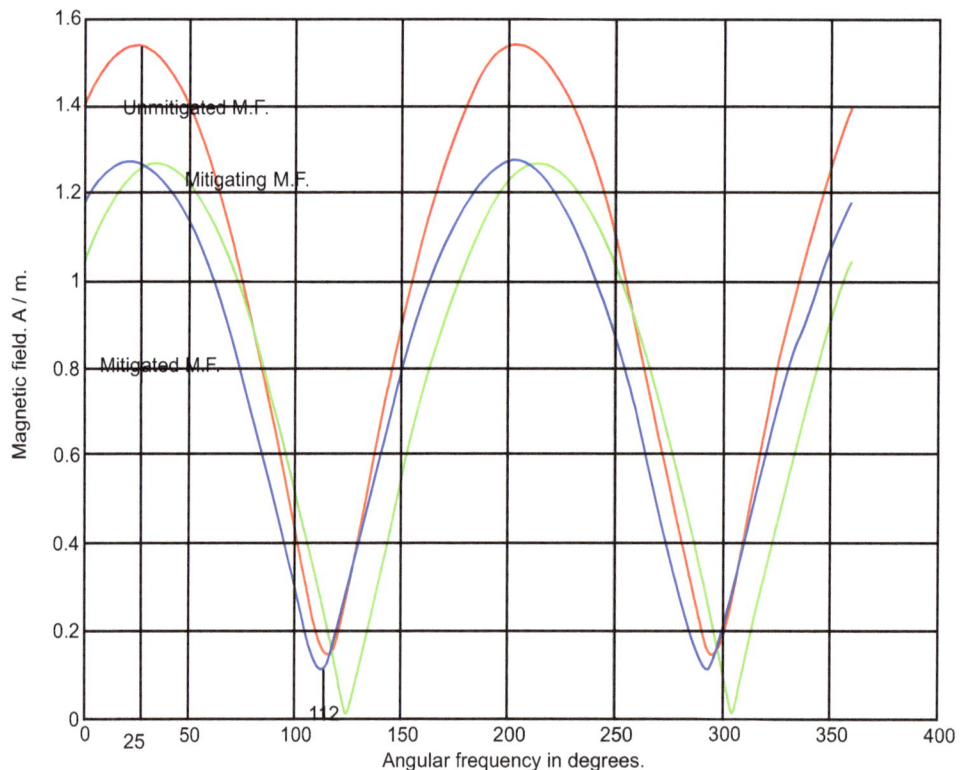

Figure 44. Characteristics of the three magnetic fields. Location, (-13.5,1).

Figure 44 shows variation of the three types of magnetic fields. As this Figure illustrates, the best mitigation, 0.1092 A / m, occurs at an angular frequency of 112 degrees.

Figure 45 illustrates unmitigated and mitigated magnetic fields with two different values of angular frequencies. In one case, angular frequency of 34.4398 degrees, responsible to produce maximum value of unmitigated magnetic field at $x_j = 9m, y = 1m$ is utilized to calculate the unmitigated magnetic fields at other six locations. In the second case, the values of angular frequencies of Table 10 responsible to produce maximum values of unmitigated magnetic fields at the corresponding locations are used. This Figure shows that the unmitigated magnetic fields in the two cases have obtained almost the same values. The mitigated field achieves different values as one moves away from location $x_j = 9m$, $y_j = 1m$ toward location $x_j = -13.5m$, $y_j = 1m$.

TABLE **12**. Relationship between the two magnetic fields. ωt varies one complete cycle

Distance from center of right-of-way. Meters	-13.5	-9	-4.5	0	4.5	9	13.5
Unmitigated M.F.	1.5377	1.9884	2.3866	2.5489	2.3866	1.9884	1.5322
Mitigated M.F.	0.1092	0.1692	0.2524	0.2958	0.2672	0.0000	0.1359

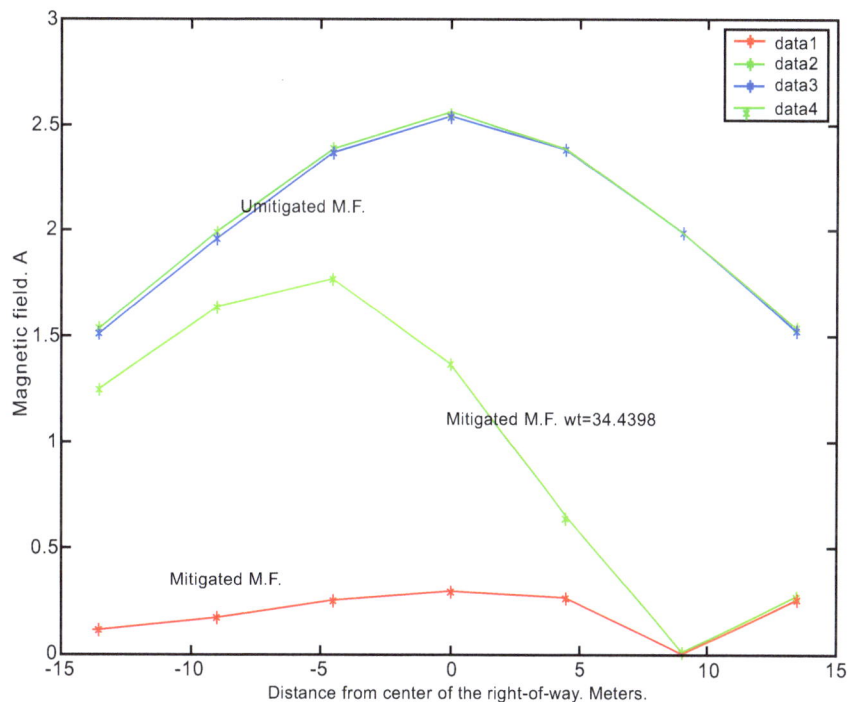

Figure 45. Comparative illustration for two values of wt. Point of consideration: (9,1).

CHAPTER 8

Bundled-Conductors vs. Single Conductor per Phase

Abstract: A comparative illustration between a flat configuration of 230 KV single conductor per phase and a bundled-conductor has been established. The obtained results which, are tabulated show that the mitigated magnetic fields remain unchanged for both the configurations.

8.1. Comparative Illustrations

Figure 46 shows a flat configuration of a 230 KV single conductor per phase transmission line. Each of the three phases carries 510 Amps. This configuration is created to establish a comparative illustration between a bundled-conductor transmission line of Figure 36 and a single conductor per phase. Let us also select the same locations, -13.5, -9, -4.5, 0, 4.5, 9, 13.5 meters.

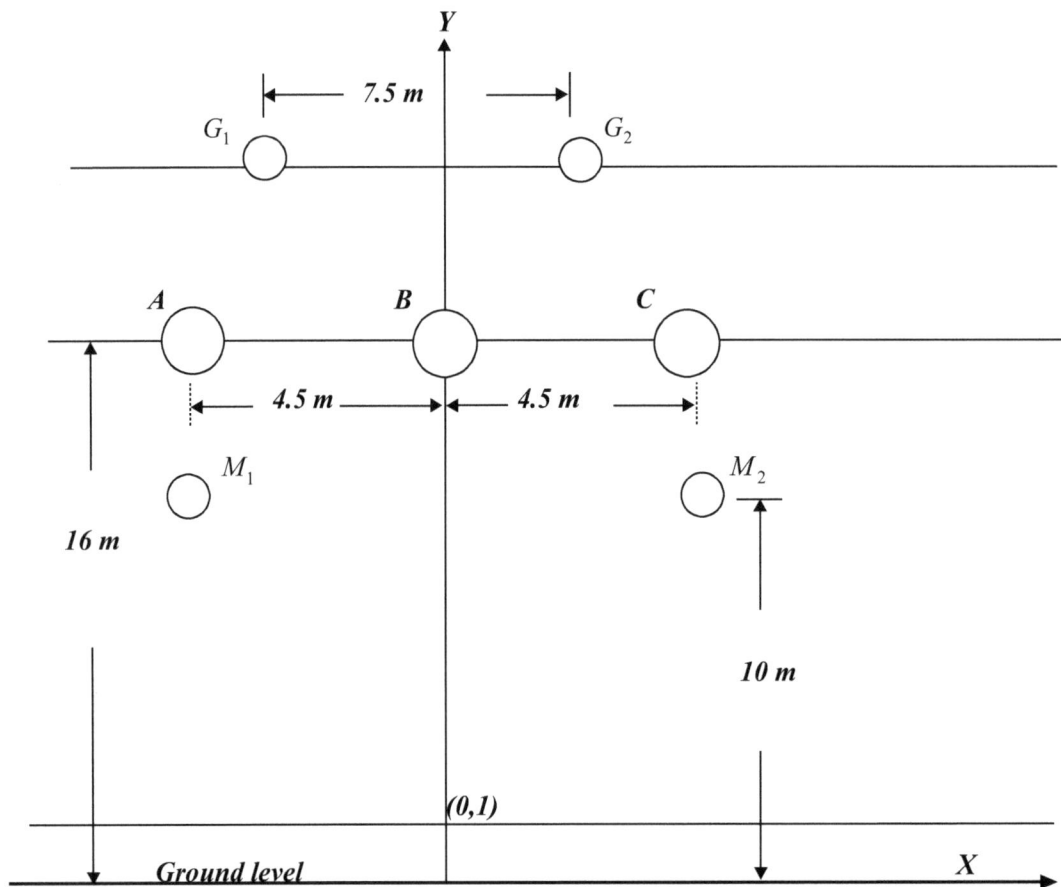

Figure 46. Flat configuration of a three phase transmission line.

Following similar procedures as was explained previously, the angular frequency responsible to generate the maximum unmitigated magnetic field at the corresponding location is calculated, the results of which are shown in Table 13.

TABLE **13**. Angular frequencies for a single conductor per phase

Distance d. Meters.	Angular frequency. Degrees
-13.5	25.0320
-9	25.5583
-4.5	27.1996
0	29.9992
4.5	32.7989
9	34.4402
13.5	34.9665

Comparing Table 13 with Table 10, similar values of angular frequencies for the corresponding distances from the center of right-of-way can be observed.

In order to establish a convincing explanation for the above phenomenon, let us scrutinize Equation (33). From this Equation, it can be written that

$$K * \sin(\omega t) = -M * \cos(\omega t)$$

therefore;

$$\tan(\omega t) = \frac{coefficient2}{coefficient1} = -\frac{M}{K}$$

Let us investigate location $x_j = -13.5m$, $y_j = 1m$ for the two configurations.

In the case of bundled conductors, implementation of Equations (34) and (35) result in achieving

K = 1.3541 + i*0.2947
M = -0.6602 + i*0.0210

And in the case of single conductor per phase,
K = 1.3752 + i*0.2887
M = 0.6685 – i*0.0255

Consequently, the ratio of $\frac{M}{K}$, generating the angular frequencies, in the two cases remain very close to each other, resulting in obtaining very close values of angular frequencies in both the cases.

8.2. Process of Mitigation

A close look at Table 12 and Table 14, it becomes obvious that even though a bundled-conductor transmission line configuration differs from a single conductor per phase configuration, but the mitigated magnetic fields remain almost unchanged.

TABLE **14**. Relationship between distance and the two magnetic fields

Distance d. Meters	-13.5	-9	-4.5	0	4.5	9	13.5
Unmitigated M.F. A/m	1.5493	2.0077	2.4139	2.5796	2.4139	2.0077	1.5493
Mitigated M.F. A/m	0.1119	0.1735	0.2579	0.3010	0.2701	0.0000	0.1368

CHAPTER 9

Auxiliary Loop – Ground Wire

Abstract: Effect of ground wire as an auxiliary mitigating loop has been investigated.

9.1. Ground Wires

Since it is not feasible to provide the high voltage transmission lines with insulators to protect them against the lightning, two conductors known as ground wires (in some cases one) are directly installed above these lines. Ground wires are grounded at frequent intervals, preferably at every pole. Ground wires cause a great reduction of dielectric stress in the air, which could be due to lightning or other atmospheric disturbances. In addition to station arresters, ground wire can act to dampen any impulses that may travel along the transmission line. Therefore, more than protecting the lines, ground wires provide a strong protection for the power station.

Since ground wires are placed along and parallel to the transmission lines, these two conductors form a loop and, consequently, a voltage is induced in this loop. Subsequently, the loop formed by the ground wires could be considered as a mitigating loop.

The loop formed by the two conductors of the ground wires may well be capable to produce hundred percent cancellation of magnetic field produced by the three-phase transmission line. In such case, the loop impedance must be thoroughly studied and optimal value of the impedance must be calculated. Such procedure, obviously, causes re-placement of the existing ground wires. Width of the loop will not cause any inconvenience, since the developed method is well applicable to any rate of loop voltage.

Installation of an auxiliary mitigating loop above the transmission line may require height of the tower to be increased. In addition, when auxiliary mitigating loop is placed above the power lines, there will surely be a mutual effect between the mitigating loop and the existing ground wires loop.

Readers are invited to thoroughly consider such mutual effect, if installation of an auxiliary mitigating loop above a three-phase transmission line is desired.

CHAPTER 10

Mitigating Loop at Ground Level

Abstract: In order to demonstrate the feasibility of the developed approach, a case when the mitigating loop is placed at the ground level is thoroughly studied. A flat configuration of 230 KV transmission line has been used and effect of mitigation at seven different locations within the right-of-way has been scrutinized and the related figures and Tables are illustrated. The loop voltage, which is the result of induced fluxes, is determined. Implementation of the previously derived equation results in achieving the value of the mitigating loop impedance. The correlation between the three types of magnetic fields has also been investigated. Finally, a comparative method when the auxiliary mitigating loop is installed above, below and at the ground level is also established and the result is shown in a Table.

10.1. Mitigating Loop at Ground Level

Figure 47 shows a flat configuration of a 230 KV transmission line with auxiliary mitigating loop $M_1 - M_2$ at the ground level.

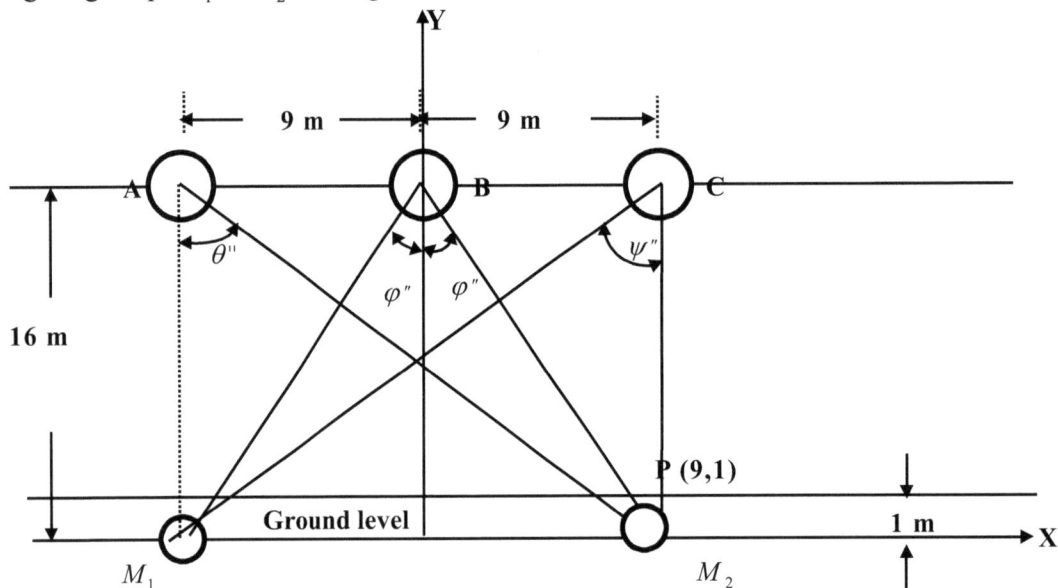

Figure 47. Geometrical position of mitigating loop $M_1 - M_2$ with respect to point P.

Even though, this Figure illustrates that centers of the loop conductors are at the ground level, but in practice and in order to safe guard the commuters against the electric shock, this loop must be buried at a slight depth.

For the purpose of demonstrating the applicability of the developed approach, seven locations such as -45, -36, -9, 0, 9, 36, and 45 meters from center of right-of-way are selected. Let the object be placed at an altitude of 1 m above the ground level. Location $x_j = 9m$, $y_j = 1m$ is selected as the point of consideration.

The usual procedures are followed to calculate the unmitigated magnetic fields at these locations.

10.2. Calculation of Loop Voltage

The three phases of the transmission line are well capable to induce fluxes in this loop. The total flux, which is the vector sum of these three fluxes are responsible to generate the loop voltage.
From Figure 47;

$$\theta' = 0$$
$$\theta'' = 0.8442$$
$$I_A = 460 \text{ Amps}$$

Substituting the above values in Equation (20), the flux induced by phase A is given by;

$$\bar{\Phi}_A = 3.7626 \text{ e-005}$$

Considering phase B;

$$\varphi' = 0.5124$$
$$\varphi'' = 0.5124$$

$$\bar{I}_B = 460[\cos(-120°) + i * \sin(-120°)]$$

Substitution of the obtained values in Equation (22) results in achieving

$$\bar{\Phi}_B = 0$$

Considering phase C;

$$\psi' = 0.8442$$
$$\psi'' = 0$$

$$\bar{I}_C = 460[\cos(120°) + i * \sin(120°)]$$

From Equation (24);

$$\bar{\Phi}_C = 1.8813e^{-005} - 3.2585e^{-005}i$$

Implementation of Equation (25) results in achieving the total flux induced in the auxiliary mitigating loop. Therefore;

$$\bar{\Phi}_T = 5.6439e^{-005} - 3.2585e^{-005}i$$

Finally, value of the loop voltage is determined by Equation (26).

$V_{loop} = 24.5685$ volts.

In order to determine the optimal value of the auxiliary loop impedance of Equation (18), Figure 48, which shows geometrical position of auxiliary mitigating loop with respect to the point of consideration P, located at $x_j = 9m$, $y_j = 1m$ is investigated.

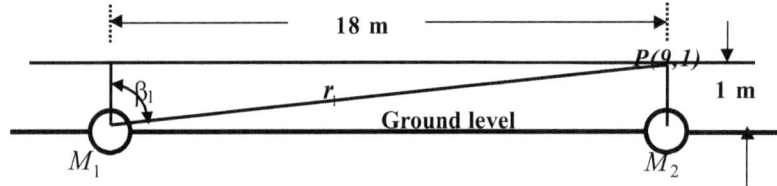

Figure 48. Position of Mitigating loop with respect to point P.

From this Figure;

$r_1 = 18.0278m$

$r_2 = 1m$

$$\cos(\beta_1) = \frac{1}{18.0278}$$

$$\sin(\beta_1) = \frac{-18}{18.0278}$$

$\cos(\beta_2) = 1$

$\sin(\beta_2) = 0$

With $\omega t = 0.5236$ rad.

$Z_m = 1.1876$ ohms

this loop impedance produces a current of 20.6869 amps.

10.3. Effect of Mitigation

Figure 49 shows position of mitigated magnetic field with respect to the corresponding unmitigated magnetic field for seven different locations, with $x_j = 9$ m, $y_j = 1m$ as the point of consideration.

As this Figure illustrates, hundred percent cancellation of unmitigated magnetic field has been achieved at the point of consideration. At all the other locations, except at $x_j = 0m$, $y_j = 1m$, remarkable reductions of unmitigated magnetic fields have been obtained.

TABLE **15**. Percentage of mitigation of magnetic fields at seven locations. Loop at ground level.

Distance from center of right-of-way. Meter	-45	-36	-9	0	9	36	45
Unmitigated M.FA / m	0.5249	0.7848	3.2874	3.7296	3.2874	0.7848	0.5249
Mitigated M.F.A / m.	0.0260	0.0532	0.1599	1.3529	0.0000	0.0395	0.0182
% of mitigation	95	93	94	64	100	95	97

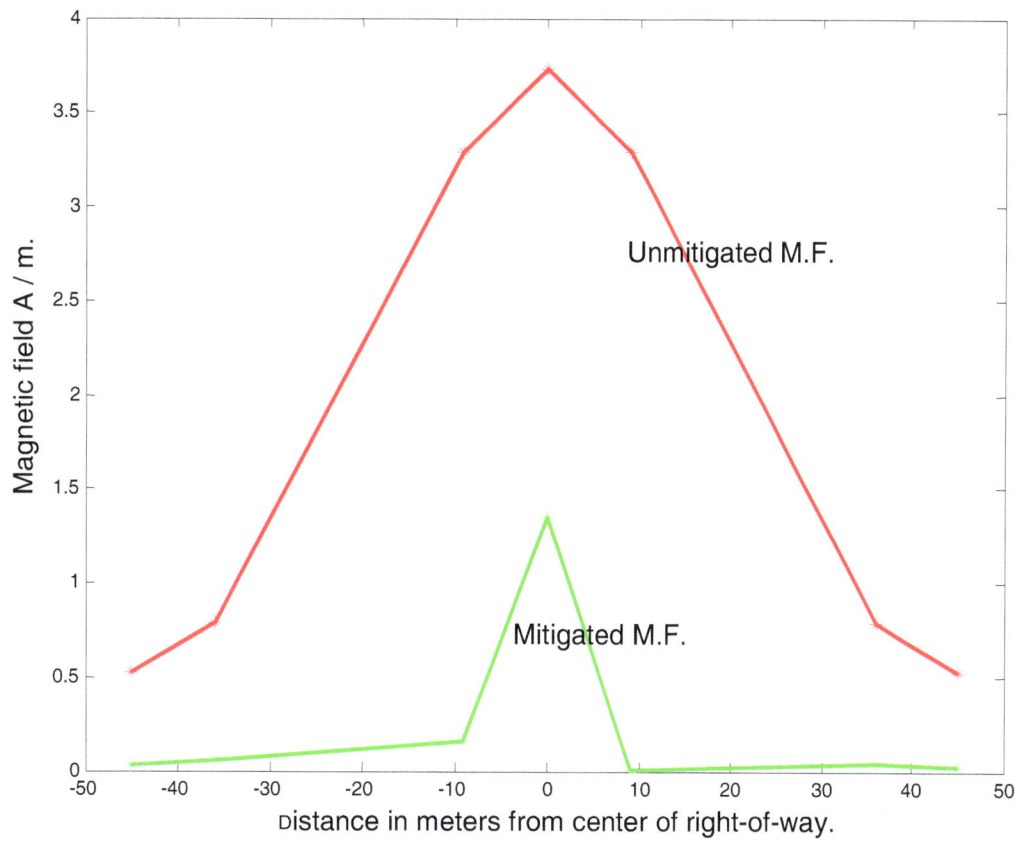

Figure 49. Relationship between unmitigated and mitigated M.F. at ground level.

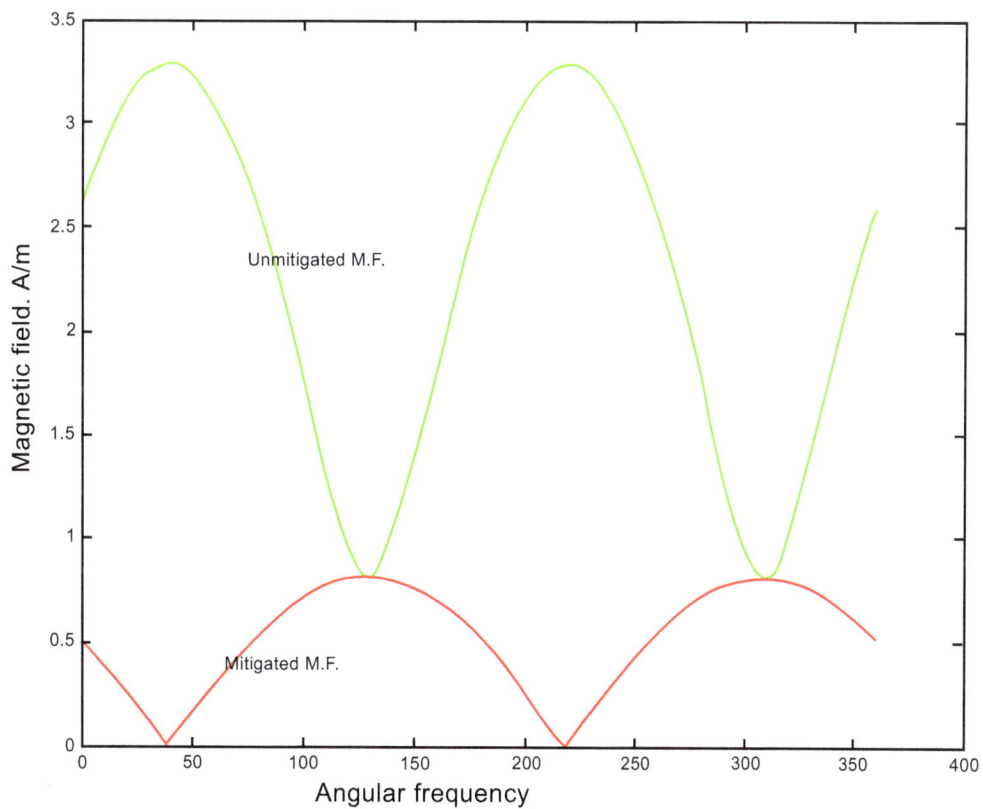

Figure 50. Position of two magnetic fields at the point of consideration.

Table 15 illuminates percentage of mitigation at seven different locations. As this Table shows, in addition to achieving hundred percent mitigation at the point of consideration, extremely high percentage of mitigations have been obtained at the other locations. With reference to this Table, it is obvious that the developed approach is well suitable to the case when the auxiliary mitigating loop is slightly buried at the ground. Such a development eases the burden of protecting the auxiliary loop against the lightning and the loop is completely safe guarded against the wind particularly, when it is gusting at a very high speed. Locating the mitigating loop at the ground level also eliminates the problems associated with the ground clearance and other phenomena such as ice loading.

Figure 51 illustrates position of unmitigated magnetic field with respect to that of mitigating magnetic field for the point of consideration $x_j = 9m$, $y_j = 1m$. As this Figure shows, the unmitigated and mitigating magnetic fields have the same magnitudes, with their orientations in the opposite directions. Consequently, the vector sum of these two fields constitutes a mitigated magnetic field of zero value. Variation of unmitigated magnetic field over one complete cycle of 360 degrees are also depicted in the same Figure.

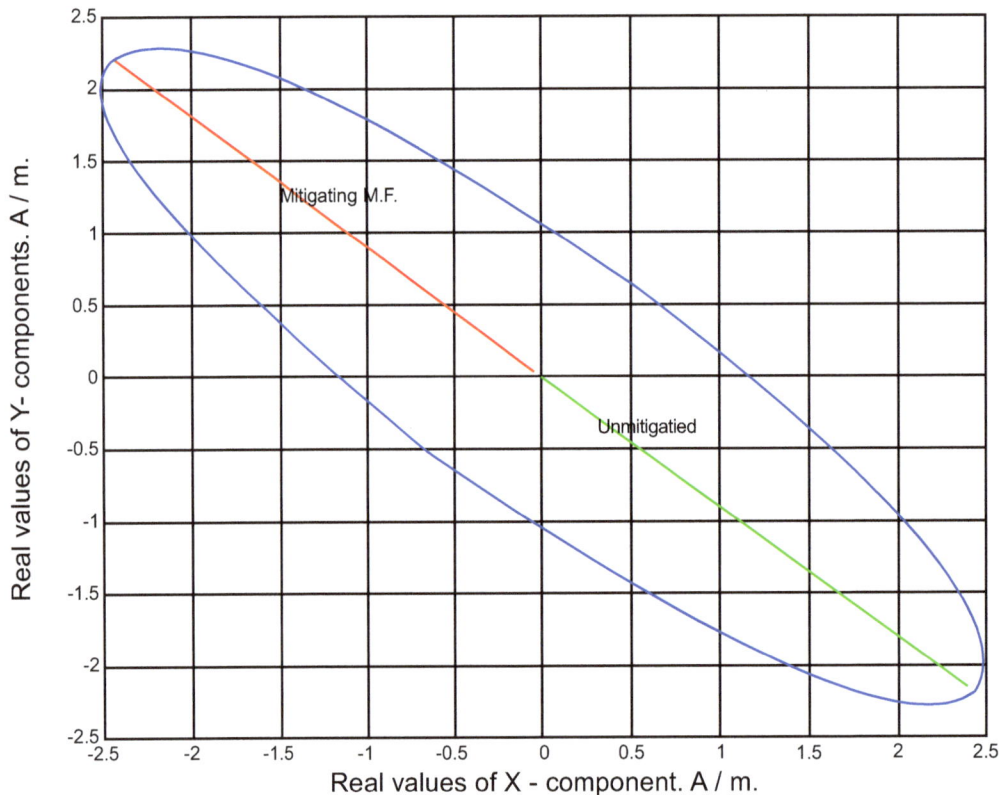

Figure 51. Position of the three magnetic fields. Loop at the ground level.

10.4. Comparative Illustrations

For further illustrating the effectiveness of position of the auxiliary mitigating loop installed

1 – above the transmission line
2 – below the transmission line
3 – at the ground level

and also to provide the readers with a quick comparison of the obtained mitigated magnetic field with installation positions of the loop, previously calculated values of the mitigated magnetic fields corresponding to these three installation positions are graphed and are depicted in Figure 52.

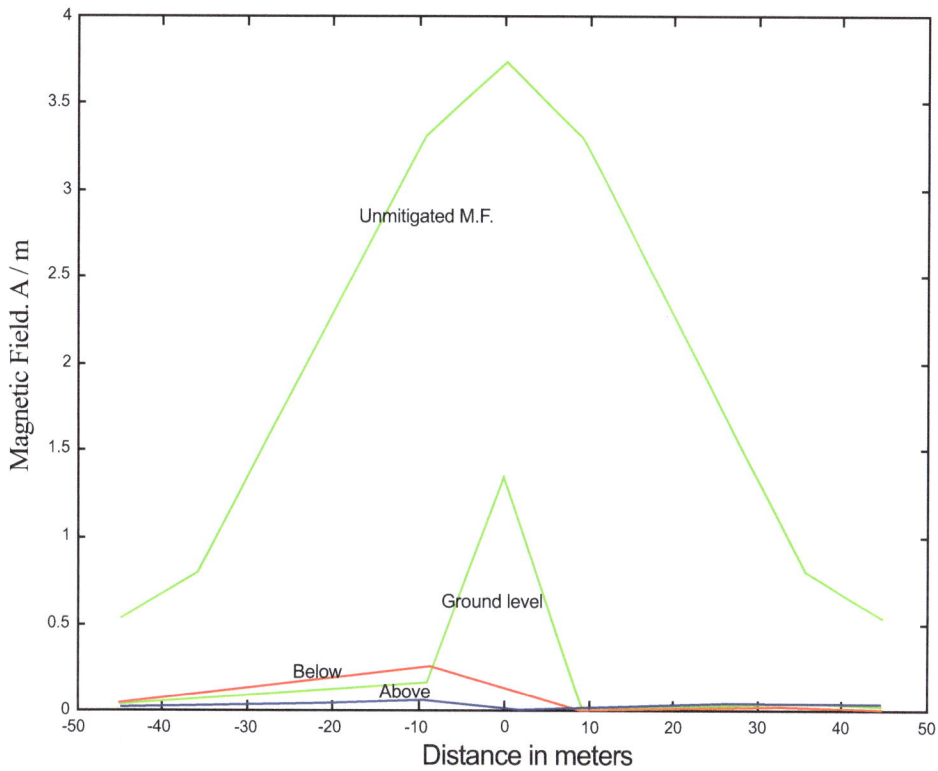

Figure 52. Characteristics of the mitigated M.F. The loop takes three different positions.

As this Figure and Table 16 show, at the point of consideration, $x_j = 9\,\mathrm{m}$, $y_j = 1\,\mathrm{m}$, value of the mitigated magnetic field is equal to zero irrelevant to the installation positions

Figure 52 and entries of Table 16 illustrate that at locations $x_j = 45$ m and $x_j = 36$ m, installation of the loop below the transmission line constitutes a better result. This Figure and the entries of Table 16 demonstrate that at other locations better mitigation is achievable, when the mitigating loop is installed above the two outer phases of the transmission line.

Since, in addition to hundred percent cancellation of the magnetic field at the point of consideration, installation of the mitigating loop at the ground level is well capable to produce between 93 and 97 percent cancellation of the generated magnetic field by the three phases of the transmission line (except at the center of the right-of-way) and since, installation of the auxiliary mitigating loop at slight depth of the ground (unlike above or below) is not associated with the problems already mentioned, it seems that installation of the loop at the ground level may be the best choice.

TABLE **16**. Mitigated magnetic fields, the loop located at three different positions

	Distance. m	-45	-36	-9	0	9	36	45
	Unmitigated	0.5249	0.7848	3.2874	3.7296	3.2874	0.7848	0.5249
Mitigated	GROUND	0.0260	0.0532	0.1599	1.3529	0.0000	0.0395	0.0182
Magnetic	BELOW	0.0481	0.0753	0.2594	0.1176	0.0000	0.0248	0.0053
Field.	ABOVE	0.0121	0.0218	0.0636	0.0107	0.0000	0.0498	0.0277

CHAPTER 11

Magnetic Field of Vertically Installed Conductors

Abstract: In this chapter, vertically arranged conductors are investigated. The total unmitigated magnetic field contributed by the three phases is calculated. The mitigating loop voltage is calculated. This calculation reveals the fact that there will be no mitigation when the auxiliary loop is installed symmetrically with respect to the Y-axis. The capability of the developed approach is well illustrated once the loop's geometrical position is changed. The fluxes induced by the three phases of the power line are well capable of producing mitigating loop voltage. Consequently, this voltage results in achieving the mitigating magnetic field, Subsequently, the mitigated magnetic field is calculated.

In order to further demonstrate the feasibility of the developed approach, the mitigating loop is placed at the ground level and the related figures are depicted.

11.1. Vertically Arranged Conductors

Figure 53 shows a three-phase transmission line whose conductors are vertically arranged. Phase A and phase B are separated by 9 meters and so are phases B and C. Current in phase A is given by;

$$\bar{I}_A = 460 * \cos(\omega t)$$

Currents in phases B and C are, as shown below;

$$\bar{I}_B = 460 * \cos(\omega t - 120°)$$
$$\bar{I}_C = 460 * \cos(\omega t + 120°)$$

This type of arrangement is well capable to produce magnetic field at any point in the space. Since the generated magnetic field has sinusoidal variation, it obtains maximum and minimum values

The angular frequency responsible to generate the maximum value of unmitigated magnetic field at the point of consideration, $x_j = 9$ m, $y_j = 1$ m, contributed by this type of arrangement can be calculated by implementing Equation (11).

From Figure 53;

$$R_A = 17.4929 \text{ m}$$

$$R_B = 25.6320m$$
$$R_C = 34.2053m$$

$$\cos(\alpha_a) = -0.8575$$
$$\sin(\alpha_a) = -0.5145$$
$$\cos(\alpha_b) = -0.9363$$
$$\sin(\alpha_b) = -0.5311$$
$$\cos(\alpha_c) = -0.9648$$
$$\sin(\alpha_c) = -0.2631$$

Substitution of the above obtained values in Equation (11), sets value of the angular frequency equal to 19.2 degrees.

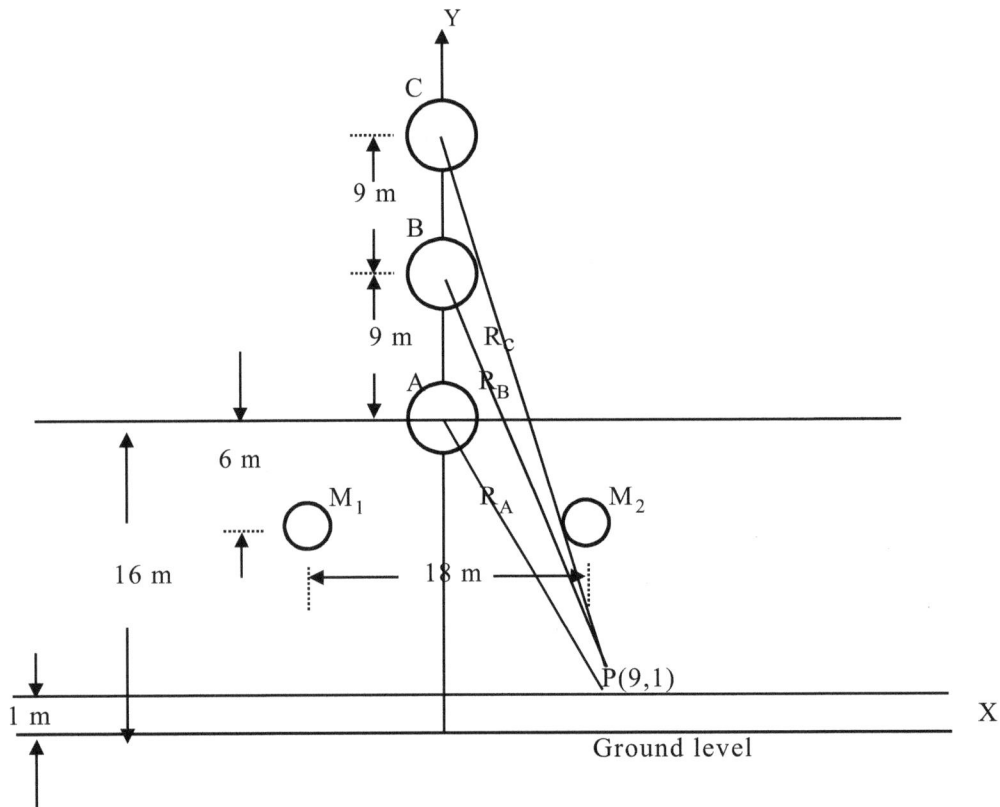

Figure 53. Vertically arranged three-phase transmission line.

11.2. Unmitigated Magnetic Field

Equation (10) is well capable to calculate the total magnetic field contributed by this type of arrangement. Substituting the already calculated values in Equation (10), sets value of the unmitigated magnetic field equal to -1.3245 - 1.4196i, a magnitude of 1.9415 A / m.

11.3. Loop Voltage

Figure 53 also illustrates that a mitigating loop $M_1 - M_2$ is installed beneath the three conductors. Due to symmetrical arrangement of this auxiliary mitigating loop with respect to Y-axis;

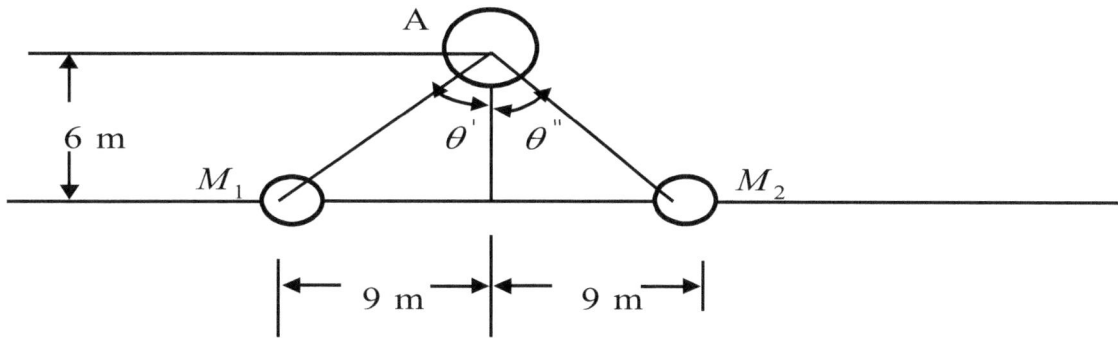

Figure 54. Geometrical location of phase A with respect to mitigating conductors.

$\theta' = \theta''$, as shown in Figure 54, sets the flux induced by phase A equal to zero. Such arrangement causes $\varphi' = \varphi''$ and $\psi' = \psi''$. Therefore, $\bar{\Phi}_B$ of Equation (22) and $\bar{\Phi}_C$ of Equation (24) would be equal to zero, respectively.

Since the total induced flux of Equation (25) is equal to zero, consequently, the mitigating loop voltage of Equation (26) would also be equal to zero. Subsequently, since there is no loop voltage, there would not be any mitigation of the magnetic field.

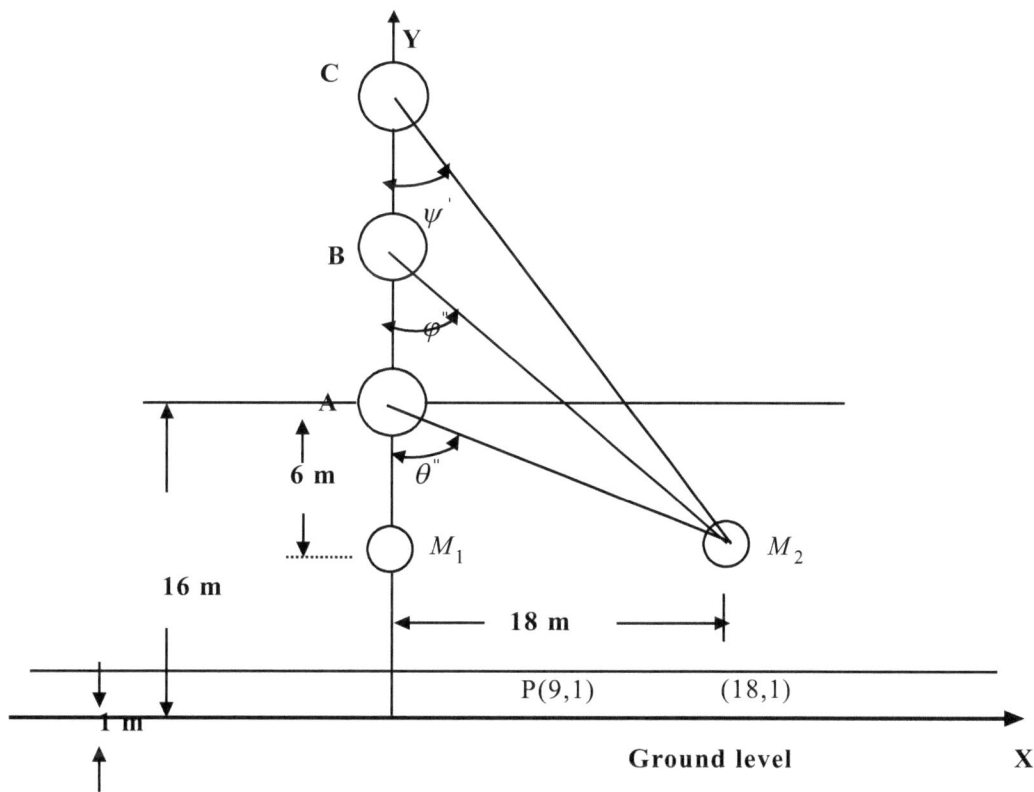

Figure 55. Geometrical position of mitigating loop with respect to the three phases.

In order to demonstrate the capability of the developed approach, let us create a slight modification to the auxiliary loop installation. As Figure 55 shows, the same loop is shifted by 9 meters toward right. Therefore, M_1 has a coordinate of $x = 0, y = 10$ and M_2 obtains a coordinate of $x = 18, y = 10$, as depicted in Figure 56.

Implementation of Equations (20), (22), (24), (25) and (26) result in achieving the mitigating loop voltage of 41.2880 volts.

Width of this auxiliary mitigating loop, obviously, influences value of the loop voltage. Let coordinate of M_1 retain a constant position of $x = 0, y = 10$. Varying the x coordinate of M_2, while its Y coordinate is kept constant at 10, influences values of sine and cosine of Equations (20), (22) and (24). Consequently, value of the loop voltage is also influenced.

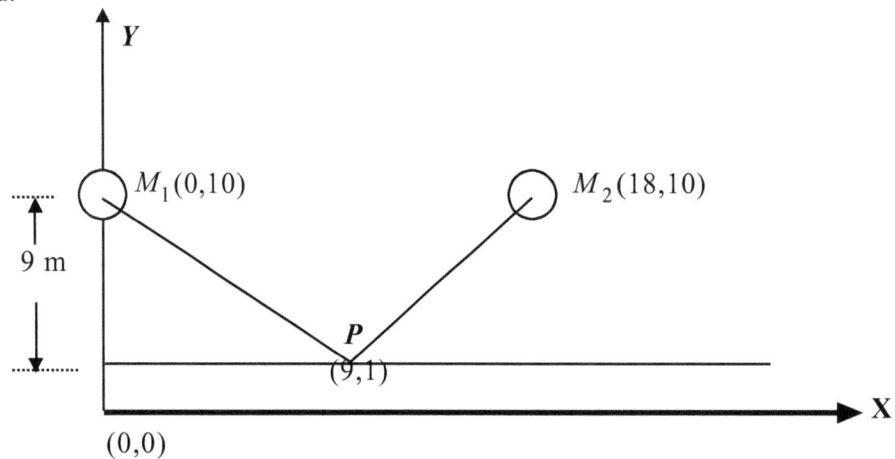

Figure 56. Geometrical location of mitigating loop with respect to point P.

The calculations have shown that value of the loop voltage increases to 72.9333 volts, when its width was increased to 36 meters.

11.4. Mitigating Field

The auxiliary mitigating loop of Figure 55 is now well capable to produce its own magnetic field. Equation (18) is well capable to establish optimal value of this loop impedance, which in turn would be responsible to generate the mitigating current along the loop.

In reference with the Equation (16), all the parameters are calculated.
From Figure 56;

$r_1 = 12.7279$
$r_2 = 12.7279$

$\cos(\beta_1) = -0.7071$
$\sin(\beta_1) = -0.7071$

$$\cos(\beta_2) = -0.7071$$

$$\sin(\beta_2) = 0.7071$$

$$\bar{I}_m = -80.2765 + 74.8987i$$

$$V_m = 41.2880 \ \text{volts}$$

Substituting the above values in Equation (16), sets value of mitigating magnetic field equal to 1.3245 + 1.4196i. Since mitigating and unmitigated magnetic fields have the same values, but their orientations are in the opposite directions, the vector sum of these two fields constitutes a mitigated magnetic field of zero value, as depicted in Figure 57.

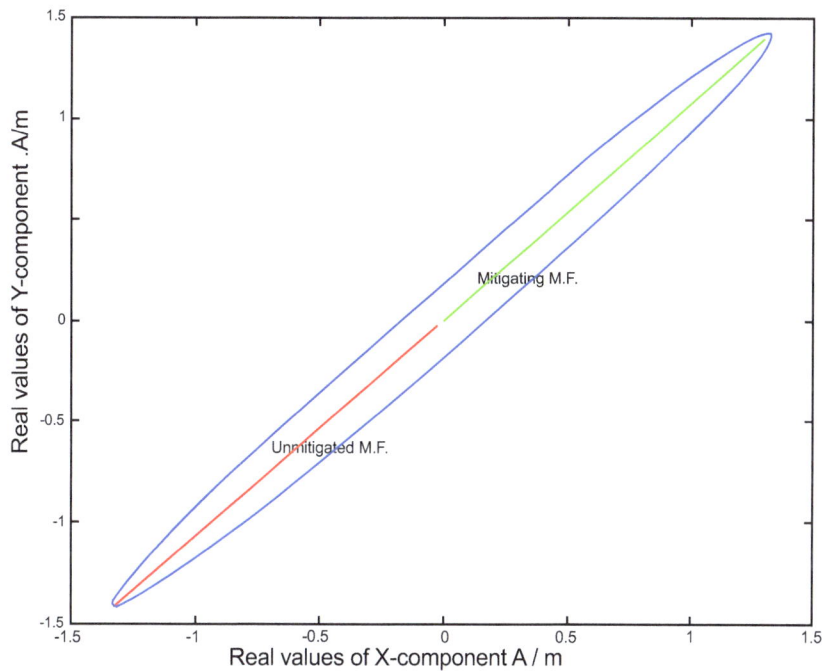

Figure 57. Variations of unmitigated magnetic field over 360 degrees.

Figure 58 shows sinusoidal variations of the unmitigated magnetic field versus angular frequency. Variation of mitigated magnetic field can also be observed from the same Figure. As depicted in Figure 58, the mitigated magnetic field achieves its zero values at angular frequencies of 19.2 degrees and 199.2 degrees, respectively.

11.5. Ground Level

In order to illuminate the feasibility of the developed approach in producing hundred percent cancellation of magnetic field contributed by the three phases of a transmission line, whose conductors are arranged vertically, when the auxiliary mitigating loop is installed at the ground level, let us scrutinize location x = 9 m, y = 1m as the point of consideration.

The auxiliary mitigating loop, $M_1 - M_2$ having width of 18 meters is installed at the ground level, as depicted in Figure 59. As this Figure shows, M_1 has coordinate of $(0,0)$ and M_2 has coordinate of $(18,0)$.

Since the same conductor arrangement of Figure 55 has been utilized, value of the angular frequency at which maximum unmitigated magnetic field of $-1.3245 - 1.4196i$ is produced remains unchanged.

These three phases are well capable to induce total flux of 4.2858e-005, resulting in a loop voltage of 16.1570 volts.

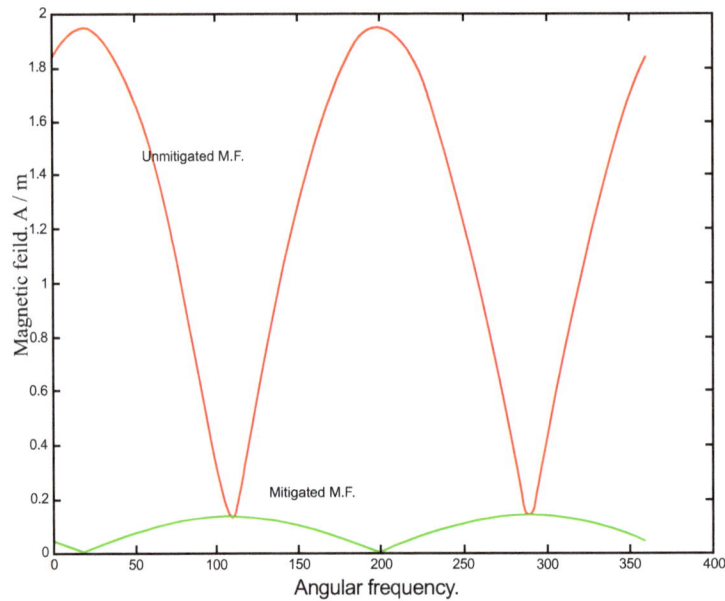

Figure 58. Characteristics of the two magnetic fields.

Implementation of Equation (18) sets value of the loop impedance equal to $-0.2126 - 0.1983i$. Consequently, a current of 55.5734 Amps would be flowing along the mitigating loop.

As Figure 59 shows;

$r_1 = 9.0554$

$r_2 = 9.0554$

$$\cos(\beta_1) = \frac{1}{r_1}$$

$$\sin(\beta_1) = \frac{-9}{r_1}$$

$$\cos(\beta_2) = \frac{1}{r_2}$$

$$\sin(\beta_2) = \frac{9}{r_2}$$

Substituting the above values in Equation (16) set value of the mitigating magnetic field equal to $1.3245 + 1.4196i$. The vector sum of unmitigated and mitigating magnetic field results in achieving the mitigated magnetic field, which is equal to zero as shown in Figure 60.

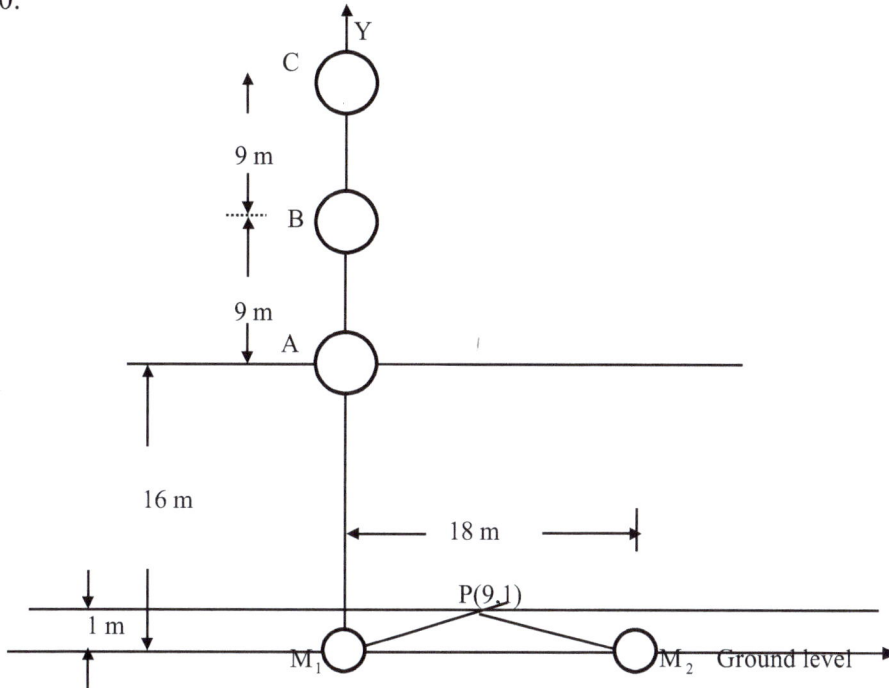

Figure 59. The loop is placed at the ground level.

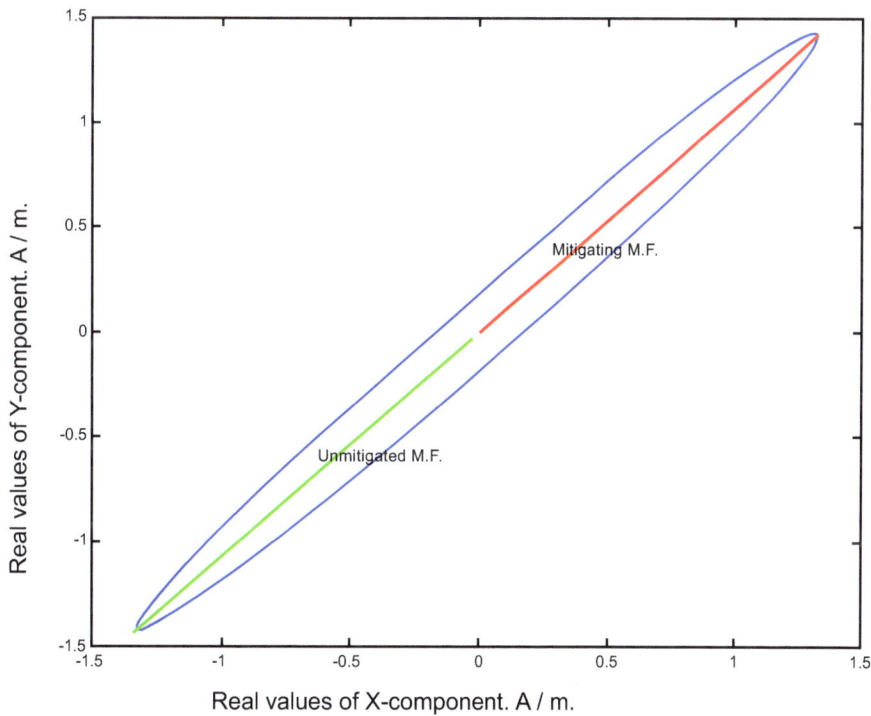

Figure 60. Relationship between mitigating and unmitigated M.F. at ground level.

REFERENCES

[1]   Chung Hi Li, Fung Chang Sung, "Association Between Occupational Exposure to Power Frequency Electromagnetic Fields and Amyotrophic Lateral Sclerosis: A Review." American Journal of Industrial Medicine, 2003.

[2]   Jukka Luutilainen, Timo Kumlin, "Occupational Magnetic Field Exposure and Melatonin: Interaction With Light-at-Night" Bioelectromagnetics, 27, 423-426 (2006).

[3]   D.E. Foliart, B.H. Pollock, *et al.*, "Magnetic field exposure and long – term survival among children with leukemia." British Journal of Cancer, (2006), 94, 161-164.

[4]   M.P. Maslanyj, T. J. Mee, *et al.*, "Investigation of the sources of residential power frequency magnetic field exposure in the UK Children Cancer Study." Journal of Radiological Protection, (2007), 41 – 58.

[5]   Kerstin Hug, Martin Roosli, *et al.*, "Magnetic field exposure and neurodegenerative diseases – recent epidemiology studies."

[6]   Z. Davanipour, E Sobel, *et al.*, "Amyotrophic Lateral Sclerosis and Occupational Exposure to Electromagnetic Fields." Bioelectromagnetics, 18:28-35, (1997).

[7]   John S. Reif, "Melatonin and Occupational Magnetic Field Exposure.", Research Information on the EMFRAPID Program, (1994 – 1998).

[8]   Hans R. Larsen, "Magnetic fields and your health." International Health News.

[9]   Robert P. Liburdy, "Environmental Magnetic Fields and Human Breast Cancer." Research Information on the EMFRAPID Program, (1994 – 1998).

[10]   A.R. Memari, "Optimal Calculation of Impedance of an Auxiliary Loop to Mitigate Magnetic Field of a Transmission Line", Power Delivery, vol. 20, no. 2, April 2005.

INDEX

www.ingramcontent.com/pod-product-compliance
Lightning Source LLC
Chambersburg PA
CBHW041720210326
41598CB00007B/718